T0135563

Mir Sayed Shah Danish

Voltage Stability in Electric Power System

A Practical Introduction

Logos Verlag Berlin

λογος

Bibliographic information published by the Deutsche Nationalbibliothek

The Deutsche Nationalbibliothek lists this publication in the Deutsche Nationalbibliografie; detailed bibliographic data are available in the Internet at http://dnb.d-nb.de .

ISBN 978-3-8325-3878-1

Logos Verlag Berlin GmbH
Comeniushof, Gubener Str. 47,
D-10243 Berlin
Germany

Tel.: +49 (0)30 / 42 85 10 90
Fax: +49 (0)30 / 42 85 10 92
http://www.logos-verlag.com

Voltage Stability
in Electric Power System
A Practical Introduction

Mir Sayed Shah Danish

Graduate Student
University of the Ryukyus
1 Senbaru, Nishihara, 903-0129
Okinawa, Japan

Statutory Statement

"Dedicated to His Parents"

___ Author

PREFACE

The understanding of power system voltage stability has become increasingly important due to day by day increase in electricity demand and liberalization policy of the electricity markets. Therefore, voltage stability has become significantly important during the past decades. The aforesaid reasons have forced the power systems to operate close to their stability limits. Both voltage stability formulation and indices are covered in this book along with an easily comprehensible manner and detailed exposition of the voltage stability indices' fundamental. However, the content of this book is considered serviceable in advanced level. The author combines his knowledge with reporting of accurate update information to illustrate the voltage stability indices and compared how to distinguish numbers of these indices in view of theirs similarity, functionality, applicability, formulation, merit, demerit, and overall performances. This book will serve as a valuable guide for the typical reader. That the readers had in mind were researchers, engineers, planners, and other professionals involved in the assessment of voltage instability in electric power system. The prerequisite for this book is suggested the basic knowledge of power system analysis and voltage stability subjects. The authorship methodology of this book was based on reference book style. Therefore, it is not recommended as a textbook.

The present edition of the book covers mainly the static voltage stability phenomenon overview as well as an introductory review of the voltage instability mechanism, voltage stability indices, and preventive measures. That followed within the framework of this book as follows: In Chapter 1, an inductive analysis of voltage stability is introduced. Afterward, an improved voltage stability assessment index based on numerical techniques is proposed in Chapter 2.

Chapter 3 is investigated on results discussion of different voltage stability indices. As a result, a novel classification for voltage stability indices is proposed. Consequently, the application of voltage stability indices are focused by two methods in Chapter 4. At first, dealt with weak buses recognition, reactive power compensation, and load shedding studies as a key solution for the voltage instability mitigation; in order to enhance the voltage stability margin. Chapter 5 is relied on the application of the voltage stability indices for load shedding purpose. Finally, the overall conclusion and main points of the entire book is briefly outlined in Chapter 6. The effectiveness of the aforesaid applications in several methodologies are demonstrated in various test systems with comparison of the recent literatures. The simulation tools used were Neplan®, MATLAB®, and PowerWorld®. Notwithstanding of the author great efforts, if any error has crept in; it would be grateful to notice as such error to amend in the forthcoming edition of this book.

DANISH Mir Sayed Shah

CONTENTS

CHAPTER 1

POWER SYSTEM VOLTAGE STABILITY

Day by day increase in electricity demand and liberalization policy of the electricity markets are persuaded the power systems to operate close to their stability limits. Nowadays, voltage stability of large interconnected power systems, due to the multiplicity of the power system operation and control components, complicated configuration and overall performance of the systems have been recognized as a complex problem for identification of the fundamental mechanism of the voltage stability.

From start of consuming electricity to now, the power systems in different countries have suffered from many blackouts related to the voltage stability problems. The 2003 was reported a severe year regarding the blackouts, which happened in United Kingdom, Denmark, Italy, and United States. For instance, the review of the United States and Canada countries power systems' blackout in 14 of August, 2003, can be a fair reminiscent of all the blackouts that were happened at that year [1] in 23 of September in Swedish and in 28 of September in Italy [2]. That around the 50 million people in Canada and United States were affected from August 14th blackout [2]. The considerable efforts of literature have been published on voltage stability analysis. These studies are covered the context of voltage stability from various aspects. So far, however, there has been terribly important questions in the research literature, to be addressed. Within the framework of this book, several research approaches are dealt altogether. The ultimate objective of this book is to address an analysis of voltage instability phenomenon, voltage stability indices, and appropriate mitigation actions of voltage instability through reactive power compensation and load shedding. Namely, the following guidelines seem to be worth pursuing in this topic:

A. An overview on voltage stability.
B. Introducing an improved voltage stability index.
C. A comparative study and classification of voltage stability indices.
D. Optimum loadability improvement of weak buses using shunt capacitors to enhance voltage stability margin.
E. Application of the improved voltage stability index for load shedding purpose.

1.1. POWER SYSTEM STABILITY

The voltage stability analysis is a part of power system stability. It seems wise to have a cursory glance at the power system stability concept, classification, and the prominence of voltage stability throughout the power system. Power system stability is classified in view of various behaviors of the systems with considering system elements and contents at different time frame. As a short classification of power system stability based on the time frame and driving force criteria (generator and load side) is shown in Figure 1.1. Power system stability is known as a key factor for secure and reliable system operation since 1920s [3, 4]. According to the [4], the formal definition of power system stability is defined that Power system stability is the ability of an electric power system, for a given initial operating condition, to regain a state of operating equilibrium after being subjected to a physical disturbance, with most system variables bounded so that practically the entire system remains intact. Briefly, from aforesaid definition can be summarized that stability is a condition of equilibrium between opposing forces [4].

However, due to the considerable extent dimensions of power stability problems that includes several categories and subcategories, it makes essential to classify the power stability problems in order to analyze specific types of the problems using and appropriate degree of detail of system representation and appropriate analytical techniques [4]. Authors in [4, 5] are categorized system stability based on the following considerations:

- The instability physical nature of the resulting mode
- The size of the disturbance
- The devices, processes, and the time span

In the next page, Figure 1.1 shows the power system stability classification in different categories based on the aforementioned criteria.

1.2. VOLTAGE STABILITY

Voltage stability is considered to be one of the keen interest of industry and research sector around the world since the power system is being operated closed to the limit. Whereas the network expansion is restricted due to many reasons such as lack of investment or serious concerns on environmental problems [6]. In general terms, voltage stability is defined the competence of a power system to continue steady voltages at all buses in the system after being subjected to a disturbance from given initial operating conditions [7]. It depends on the system ability to restore again to the equilibrium condition with respect to load demand and supply variation in the system, as well as the power system components various performance due different conditions of operation. From various definitions related to power voltage stability, the consensus conceptual definition summarizes in view of the power system load demand and supply equilibrium that power system voltage stability is restoration state of a system (to normal steady condition) after being subjected to a disturbance.

Due to the tendency of increasing reactive power demand on the system because of a combination of events and system conditions, additional reactive power demand may lead to voltage collapse, causing a major blackout of part or all of the system. It is extremely important to ensure the stability of the system in such situations. Voltage stability depends on how variations in reactive, as well as active power in the load area, affect the voltages at the load buses.

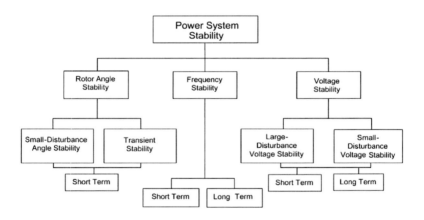

Figure 1.1   Classification of power system stability [4].

1.3.   GLOBAL BLACKOUT INCIDENCES

From the start of consuming electricity to now, the power systems in different countries have suffered from many blackouts related to the voltage stability problems in the systems. The 2003 was reported a severe year

4

regarding the blackouts, which happened in United Kingdom, Denmark, Italy, and United States. For instance, the review of causes of the United States and Canada countries power system blackout in 14 of August, 2003, can be a fair reminiscent of all the blackouts that were happened at that year [1] in 23 of September in Swedish and in 28 of September in Italy [2]. Around the 50 million people in Canada and United State were affected from August 14th blackout [2].

Finally in 2004, the North American Electric Reliability Council (NERC) was recommended some reforms in order to improve the system overall reliability. These recommendation were covered three main parts of power system from generation to distribution and utilization such as balance in power demand and generation continuously, maintain the scheduled voltages with considering the reactive power balance in the system, timely monitor the thermal and stability limits of the transmission lines, preparation for emergencies, and so on. The reasons were caused the blackout were mainly reported the deficiencies in specific practices, inadequate system understanding from system, lack of timely preventive action of cascading, equipments, lack of adequate real-time data, interference of various organizations as well as lack of reactive power in the system.

The lessons learned from Swedish disturbance or the involved factors to long-term voltage instability that was caused blackout, are detailed in [8] as following:

- Stressed power system due to heavy loading the system.
- Insufficient local reactive power supply due to current limiters of the generators
- None distinguished of load characteristics at low voltage magnitudes
- Tap changer respond to demand-side
- Improper operation of the relays in the system

On July 23, 1987, Tokyo area was experienced a massive blackout that was a consequence of long-term voltage instability. In this blackout, 2.8

5

million consumers (8GW) for 3.35 hours was affected [9]. The reasons behind the Tokyo Electric Power Company (TEPCO) network on July 23, 1987, was reported as follows [9, 10]:

- Inadequate reactive power planning and at the same time imbalance behavior of automatic capacitor bank switching with the load increase speed of 400MW/min (incompetent demand forecasting characteristics).
- Insufficient and timely monitoring of the 500 kV network due to lack of visualization control facilities.
- Lack of pre-defined strategies against probable voltage instability at that time, some of the operators didn't have an adequate comprehension about the phenomenon and its countermeasures.
- Exceed the maximum allowable peak demand that was forecasted. The TEPCO was forecasted 38.5 MW based on 33C, but due to change in weather and demand; it was increased to 39 MW on 34C at 8 am to 40 MW on 36C at 11 a.m.
- Most of the transformer in the TEPCO's transmission networks were equipped with on-load tap-changer that is controlled by Reactive Power (Q) Controller (VQC), which was known unwanted control.
- Incoordinate distribution of power plants, which were concentrated in the eastern and northern end of the TEPCO network and around Tokyo bay.
- As a result, the lessons were learned from the blackouts are listed as below:
- Objective evaluation of the high and flat voltage profile through pre-prepared for critical load seasons and times.
- Decided for the preventive actions in order to recover such incidence.
- Sufficient monitoring and evaluation of the system.

1.4. VOLTAGE STABILITY CLASSIFICATION

The exact classification of voltage stability indices seems perplexing. Therefore, it is arduous to classify a stability phoneme absolutely to the specific class. However, that the voltage stability analysis in use are very different and for convenience various classifications are proposed in the literature based on the general closeness of the behaviors characteristics. In general, each analysis has its own significance and restriction from power system operation behavior analysis point view (static, dynamic, quasi-steady state, and transient analysis) as well as from practicability in real world application for various purposes such as, voltage stability margin enhancement, weak buses/areas and permissible loading capability identification, optimal placement of distributed generation, reactive power dispatch, power management, and etc. in a power system.

1.4.1. Types of Voltage Stability Analysis

The voltage stability phenomenon customarily classified in steady state, quasi-steady state, dynamic, semi-dynamic, and transient analyzes. Obviously, there are mainly relied on two main approaches of voltage stability analysis static and dynamic. Beside of the static aspect of voltage stability, still there is required the dynamic mechanism of a real system to be investigated. As defined that a power system is a high-order multivariable nonlinear system; that operates in a constantly changing environment with a dynamic response influenced by a wide array of devices [11]. Therefore, both static and dynamic behaviors of the system involve in voltage stability [12]. Moreover, distinguish between dynamic and static aspect of the voltage stability is a common misconception problem.

Voltage stability static analysis can be solved with just algebraic equations, so it is computationally much efficient and easy compared the dynamic analysis. Static analysis captures snapshots of system conditions at various

7

time frames along the time-domain trajectory [11]. Static aspect of voltage stability defines maximal admissible load margin beyond which a load flow solution no longer exists. Dynamic analysis expresses in the form of non-linear differential and algebraic equations.

Manifestly, there are two aims behind the steady state voltage stability analysis, first, to estimate how far a given operating point is far away from stability limit (stability margin) and the second, to identify the critical loading condition.

The quasi-steady state (QSS) analysis consists in simulating the long-term dynamics with the short-term dynamics replaced by their equilibrium equations. QSS long-term simulation offers an interesting compromise between the efficiency of static methods and the advantages of time-domain methods [11]. Dynamic analysis expresses in the form of non-linear differential and algebraic equations.

In the semi-dynamic analysis, the dynamic model was considered by the steady state operation based on the stable equilibrium operating point of the power system [13, 14]. Study of the voltage stability phenomenon due to many contributing factors to the problem has dynamic and transient behavior in the nature. In view of the preferability of the voltage stability assessment techniques; primarily, the computation time, technique ease, storage requirement, data acquisition, online or offline applications, and so on, can be a benchmark for relying on a technique. Perhaps, the simulations and acquirability of the results of the dynamic and transient voltage stability analyzes are not readily possible with considering the aforementioned factors. Therefore, the steady state voltage stability analysis can pave the ground for preliminary stability analysis as a prerequisite tool [2].

In general, there are difference of opinion of scholars on the exact definition of this classification. However, the consensus with considering the measure of disturbance and its time interval in the system on the subject is defined at first in two categories, in which each category includes two parts. Shown in Figure 1.2.

- Classification In view of Voltage Instability Occurrence Time
  A. Short-term voltage stability
  B. Classical voltage instability
  C. Long–term voltage stability
- Classification In view of type of the disturbance
  A. Small disturbance voltage stability
  B. Large disturbance voltage stability

1.4.2. Voltage Instability Classification In view of Time Occurrence

1.4.2.1. Short-term Voltage Stability

Short-term voltage stability is referred to a short timeframe in order of few seconds or fractions of a second as a fast phenomenon and also sometimes known as transient voltage stability. In other word, the short-term voltage instability is known transient voltage instability as well. Voltage instability is a result of inadequate reactive power in the system similar to the other types. This type of voltage instability affected from intensity of the inductive load (induction motors), which consume an enormous amount of Mvar in the system. In general, there is difference opinions in voltage stability classification. These instability problems are often happening due to the fast response of power system elements such as controllers, generators' automatic voltage regulator (AVR), power electronics devices (FACTS) [11].

This type of stability involves dynamics of fast acting load component such as induction motors, electronically controlled loads and HVDC converters with interesting study period of several seconds. Time domain transient stability is dealt with rotor angle of synchronous machine as well as dynamic performance of induction motors and another control component in the system [18].

1.4.2.2. Classical Voltage Instability

This type of instability is attributed to a deficiency in reactive power in which the system cannot recover the disturbance. The interconnected system mostly experienced this type of instability. This type of instability may take from 1 to 5 minutes [17].

1.4.2.3. Long-term Voltage Stability

Long-term voltage stability involved slower acting equipment such as load recovery by the action of on-load tap changer or through load self-restoration and delayed corrective control actions such as shunt compensation switching or load shedding [11]. In other word, the long-term voltage stability explains the dynamic performance of the system through acting of power system equipments in the system such as tap-changing transformers, thermostatically controlled loads and generator current limiters [7]. The long-term voltage stability from reactive power view point is referred to the function of transmission lines to connect the source (generators) to the demand (consuming side) in order to transfer adequate reactive power to the points where the Mvar is needed. The study period of interest may extend to several or many minutes. The modeling of long-term voltage stability requires consideration of transformer tap changers, characteristics of static loads, manual control actions of operators and automatic generation control. In long-term stability it is essential to distinguish between two types of stability problems which are; voltage problem due to system configuration and system dynamic behavior, and frequency problem in the system due to generation and load imbalance. Table 1.1 Summarizes and provides a concise overview of the three types disturbance.

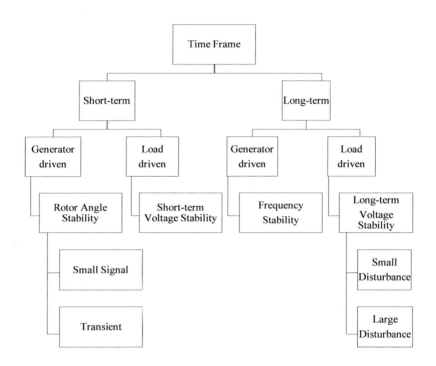

Figure 1.2   Classification of power system voltage stability [17].

1.4.3. Voltage Instability Classification In view of Disturbance Occurrence

1.4.3.1. Small Disturbance Voltage Stability

Small disturbance voltage stability phenomenon is introduced the behavior of a power system under small disturbance voltage stability which arise due to interaction of all the load characteristics (steadily increase or sudden mismatch), control and protection devices for a portion of a second in instant of time.

1.4.3.2. Large Disturbance Voltage Stability

Large disturbance voltage stability is the behavior of stability in a power system usually appears due to large disturbance in the system such as significant faults, interconnected contingencies, lack of demand and supply meet (load characteristics), and unallowable losses in the system. The study period of interest may extend from a few second to tens of minutes.

Figure 1.3   Power transmission tower (2013, Okinawa, Japan; by Author).

Table 1.1   Voltage instability types in different time frames [17].

No	Type	Cause of incident	Time frames
1	Long-term	Slowly use up of reactive reserves, and No outage.	Several minutes to several hours
2	Classical	Key outage leads to reactive power shortage.	One to five minutes
3	Short-term	Induction motor stalling leads to reactive power shortages.	Five to fifteen seconds

1.5.   CUSTOMARY CONCEPTUAL VOLTAGE STABILITY INDICATORS

Customary the P-V, Q-V, and P-Q-V fundamentals smooth the path of understanding the voltage stability behavior due to static power flow simulation. But, at the collapse point, there is no real solution for power flow. Because of single point beyond that is beyond the bounds of possibility to solve the power flow [18].

In this way, the voltage collapse proximity indexes are originate in order to predict the distance between the particular operating point and probable voltage collapse point. Also, that is known loadability indicator or maximum loadability point approach. From load flow viewpoint, the indicators of the dynamic and steady-state analysis are included in the following terms [2]. That the proposed acceptable voltage profile across the system is reference.

- Maximum power transfer limit (P-V curve)
- Reactive power capability (Q-V curve)
- Voltage instability (load flow) proximity index

1.5.1. Maximum Power Transfer Limit (P-V Curve)

The P-V curve, often use for voltage security evaluation [2], that determine the distance between the operating point (real power) and instability point. The P-V curve indicates that as the reactive power demand increases, the bus voltage steadily decreases. This condition continues to the knee or maximum power transfer limit (also is called critical point or value, and voltage stability and instability boundary), any further increase in power transfer (load demand) causes a rapid decrease in voltage. Figure 1.4 shows the 3-bus system voltage profile in respect to the change in system load. PV curve in collaboration with the PQ curve is typically used as an analysis tool to study the voltage collapse in the power system [17].

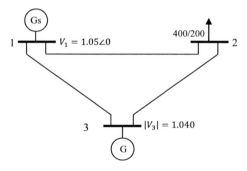

Figure 1.4   FGL 3-bus system [19].

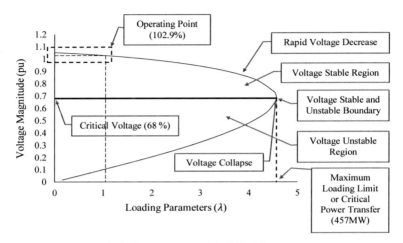

Figure 1.5 The 3rd bus P-V curve of the FGL 3-bus

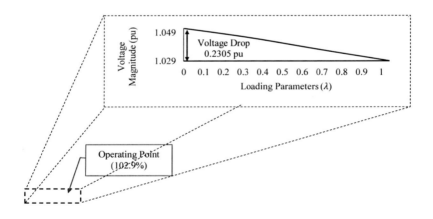

Figure 1.6 Operating point illustration.

1.5.2. Reactive Power Capability (Q-V Curve)

The Q-V curve expounds the sensitivity and variation relationship between network voltage profile and network injected reactive power. In other word, Q-V curve can predict the Mvar margin before reaching minimum voltage limit that helps to size the reactive power compensations (shunt capacitor or static var) [20].

The Q-V curve originates the power margin concept in which it study under active and reactive power margins. The incremental increase relationship between the system injected reactive power, and its voltage magnitude are depicted in Figure 1.8.

1.5.3. Voltage Instability (Load Flow) Proximity Index

The P-V cure does not include the reactive power component; therefore, there is a need to consider the third aspect as well. Numbers of trajectories to voltage collapse can be derived. Hence, many voltage collapse indexes are proposed in which to find the system parameters that define the system close to instability. Usually, these indexes are derived based on the sensitivity of the reactive power generation with respect to the load parameters. At unstable condition, the sensitivity is the impressive, small increase in load causes the abrupt increase in reactive power absorption in the system. In other word, the rate of change of reactive power absorption and load increase leads but to infinity rate of change.

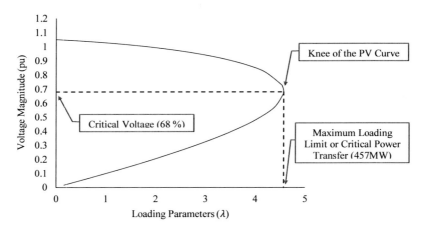

Figure 1.7 The Loading margin at bus 3 of the FGL 3-bus system.

Just rely on either P-V or V-Q analysis is not quite enough to judge that the power system voltage stability condition and assesses the proximity to voltage collapse point. At the same time, coordination of both P-V and V-Q including P-Q analysis can be count a milestone of the results to be investigated. Here, the P-V and V-Q illustrates the stability margin and P-Q introduce the sensitivity condition of the power system [21]. The specific function of sensitivity analysis is known significant for power system voltage stability analysis beyond the other methods, which predict the critical point. It is important to investigate how this critical point is affected by changing the system conditions [22].

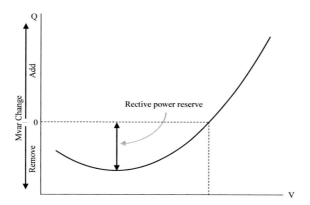

Figure 1.8 Q-V Curve.

Sensitivity analysis can be a useful tool for determining, weak bus, active and reactive power losses, and the reactive power margin (MVAR distant to voltage collapse point) [22]. A system is voltage stable if V-Q sensitivity is positive for every bus, and unstable if V-Q sensitivity is negative for at least one bus [16].

1.6.   VOLTAGE STABILITY FORMULATION

Since the electricity was consumed as key for convenience life, all along the three main parts of a power system, from power station (power generation) to consumers (power consumption or load) through transmission and distribution networks have influenced the voltage stability problems. In this study, the voltage stability assessments are studied in respect to the two a) load, and b) transmission aspects, mainly from static behavior point view. This section introduces the basic formulation of voltage stability and overview the system key parameters, which involve in voltage stability of the power system.

For proper understanding, two bus system is customary supposed to illustrate more concise concept of voltage stability. System variables-based and Jacobian matrix-based voltage stability analysis are often formulated with 2-bus system. Which are based on power flow analysis and the Jacobian matrix [2]. The 2-bus system's single phase equivalent circuit of a generator connected to load bus through transmission line is shown in Figure 1.9.

Many efforts are focused on the analysis of two-bus system in order to easy analytical derivation and provide insight into the problem that are adapted to large system of arbitrary complexity.

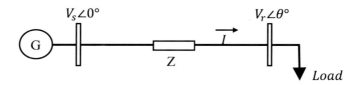

Figure 1.9 Customary 2-bus system.

where, Z is the series impedance, and V_s and V_r denote the source bus voltage and the load bus voltage magnitudes respectively. The two bus system in Figure 1.9 contains a slack bus (generator bus) and a load bus, which are connected through a transmission line. Which, these assumptions are considered:

- Transmission line is lossless.
- The voltage magnitude at both buses is considered constant.
- The maximum power transfer is calculated at $\theta = 90°$.

From equivalent basic circuit of 2-bus power transmission system, by using Kirchhoff's voltage law (KVL) the following mathematical expressions can be written:

$$V_s - V_r - ZI = 0 \qquad (1.1)$$

From Kirchhoff's Voltage Law (KVL) we immediately find the current in the as circuit as:

$$I = \frac{V_s - V_r}{Z} \qquad (1.2)$$

$$Z = jX \qquad (1.3)$$

From power transfer formula:

$$P + jQ = S \qquad (1.4)$$

$$S = V_r I^* \qquad (1.5)$$

$$S = V_r \left(\frac{V_s - V_r}{jX}\right)^* \qquad (1.6)$$

$$S = V_r \angle \theta \left(\frac{V_s \angle 0 - V_r \angle \theta}{jX} \right)^* \tag{1.7}$$

$$S = V_r \angle \theta \left(\frac{V_s \angle 0 - V_r \angle - \theta}{-jX} \right) \tag{1.8}$$

$$S = V_r (\cos \theta + j \sin \theta) \left[\frac{V_s - V_r (\cos(-\theta) + j \sin(-\theta))}{-jX} \right] \tag{1.9}$$

$$S = V_r \cos \theta + j V_r \sin \theta \left[\frac{V_s - V_r \cos \theta + j V_r \sin \theta}{-jX} \right] \tag{1.10}$$

$$S = V_r \cos \theta + j V_r \sin \theta \left[\frac{j V_s - j V_r \cos \theta + j^2 V_r \sin \theta}{X} \right] \tag{1.11}$$

$$S = -\underbrace{\frac{V_s V_r}{X} \sin \theta}_{P} - j \underbrace{\left(\frac{V_s V_r}{X} \cos \theta - \frac{V_r^2}{X} \right)}_{Q} \tag{1.12}$$

$$P = -\frac{V_s V_r}{X} \sin \theta \;\Rightarrow\; \sin \theta = -\frac{PX}{V_s V_r} \tag{1.13}$$

$$Q = \frac{V_s V_r}{X} \cos \theta - \frac{V_r^2}{X} \;\Rightarrow\; \cos \theta = \frac{QX + V_r^2}{V_s V_r} \tag{1.14}$$

From these equations, can be observed that the real power transfer directly affected from the change in power angle. The $1 = \sin^2 \theta + \cos^2 \theta$ is used in order to find the relations between variables as following:

$$1 = \left(-\frac{PX}{V_s V_r}\right)^2 + \left(\frac{QX + V_r^{\,2}}{V_s V_r}\right)^2 \tag{1.15}$$

with transforming the above equation we can obtain:

$$(PX)^2 + (QX)^2 + 2QXV_r^2 + V_r^4 - (V_s V_r)^2 = 0 \tag{1.16}$$

$$X^2(P^2 + Q^2) + V_r^2(2QX - V_s^2) + V_r^4 = 0 \tag{1.17}$$

$$\frac{V_r^4}{V_s^4} + \frac{V_r^2}{V_s^4}(2QX - V_s^2) + \frac{X^2}{V_s^4}(P^2 + Q^2) = 0 \tag{1.18}$$

After the normalization of the variables, $= \frac{V_r}{V_s}$, $p = \frac{PX}{V_s^2}$, and $q = \frac{QX}{V_s^2}$; and using the identities through form transformation the following equation can be generated:

$$v^4 + v^2(2q - 1) + p^2 + q^2 = 0 \tag{1.19}$$

$$v^4 + v^2(2p \tan \phi - 1) + p^2 + (p \tan \phi)^2 = 0 \tag{1.20}$$

$$v^4 + v^2(2p \tan \phi - 1) + p^2 + p^2(\sec^2 \phi - 1) = 0 \tag{1.21}$$

$$v^4 + v^2(2p \tan \phi - 1) + p^2 \sec^2 \phi = 0 \tag{1.22}$$

By help of the Muhammad ibn Musa al-Khwarizmi method, the quadratic equation has four solutions, which the two solutions satisfy in order to define the high and low voltages margin.

$$v^2 = -(2p \tan \phi - 1) \pm \frac{\sqrt{(2p \tan \phi - 1)^2 - 4p^2 \sec^2 \phi}}{2} \tag{1.23}$$

$$v^2 = -(2p \tan \phi - 1)$$
$$\pm \frac{\sqrt{(2p \tan \phi - 1)^2 - 4p^2(\tan^2\phi + 1)}}{2} \tag{1.24}$$

where, $V_s \angle 0°$ and $V_r \angle \theta°$ are voltages at the sending end and the receiving end buses. Z is the transmission line impedance. Obviously seem that voltage at load bus (V_r) is directly depend to Supply Voltage (V_s) with a little differ of transmission line characteristics. The maximum power transfer take place when $\theta = 90°$. For better illustration the relation between θ and power transfer (P) is shown in Figure 1.11, in Section 1.8.1.

Note: The bolt notation defines a vector quantity. The list of identities, which are used in order to simplify the equations are provided in Annex 1.2.

1.7. VOLTAGE STABILITY FROM LOAD ASPECT

1.7.1. Voltage Dependence of Load

Based on the various load devices, there are several load models, which are divided mainly into two categories:

A. Static load model
B. Dynamic load model

1.7.1.1. Static Load Model

The static load model of constant power, constant current and constant impedance, known as ZIP [23].

1.7.1.2. Dynamic Load Model

In a real power system, the demand aggregates from different types of loads (inductive and resistive loads) such as residential, commercial, industrial etc.

1.7.2. On Load Tap Changers (OLTC)

OLTC temporarily improves the distribution or sub-transmission voltage while, it does not change the transmission voltage. There is still lack of reactive power reserve. Used in order to keep the operating voltage of the regulated bust within some acceptable limit.

1.7.3.1. Effect of OLTC Transformer on Voltage Stability

It is interesting to contemplate that OLTC Transformer operation ensure nominal or near to the nominal voltage at secondary when it has inevitably impacted on the primary side. Despite this, the lack of reactive power instrumental in heavy load in secondary that grounds for voltage instability (gradually collapse). OLTC Transformer restores the desired voltage value in secondary but at the same time increase the risk of voltage collapse. The OLTC in real-life application is confined to small range in order to solve voltage instability problem as remedial action.

1.8.   VOLTAGE STABILITY FROM TRANSMISSION ASPECT

Mainly, the instability phenomenon in transmission system is considered from two aspects; the capability of transmission (maximum power that can be delivered to loads) and the correlative relation between system voltage and load [18]. Transmission systems also have a lion share in power system stability that the voltage stability is directly affected from this aspect. The thermal or stability constrains determine the transmission networks capacity [22].

1.8.1. Power Transfer Limit

As pointed out previously, power deliverable limit can be the result of instability in systems. So, in this section is dedicated to focusing on the determination of the maximum power delivery at the receiving end of the simple 2-bus system. Voltage instability stems from the attempt of load dynamics to restore power consumption beyond the capability of the combined transmission and generation system.

According to the [17], the power transfer capability is the amount of MW that can be transferred across a transmission path while ensuring that a single most severe contingency will not result in unacceptable transmission system consequences. That NERC has defined two terms to assist with determining and marketing transmission capacity:

- The total transfer capability is the transmission path's maximum safe MW transfer limit.
- The available transfer capability is the portion of the total transfer capability still available for commercial usage.

The aim of this study is to determine the maximum power to be transfer in a simple 2-bus system, in order to derive the voltage collapse point that associated with maximum permissible power at specific critical voltage. The 2-bus representation, by definition, the voltage magnitude and frequency are constant at the system. Furthermore, there is assumed 3-phase balanced steady-state sinusoidal operation condition (that the single phase is illustrated in Figure 1.10), ideal voltage source   E transmission line with series resistance R and reactance, neglected line shunt capacitance (for simplicity) as given by the customary pi-equivalent. From Figure 1.10, the transmission line power transfer limit is driven as [24]:

$$P_s = \frac{E_s E_r}{V} \sin \delta = P_{max} \sin \delta \qquad (1.25)$$

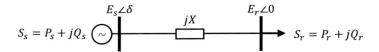

Figure 1.10   Elementary model for calculation of power transfer limit.

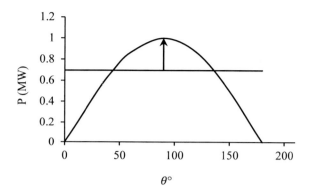

Figure 1.11 Sample variation of real power vs. power angle.

When the load impedance is equal to the complex conjugate of the transmission impedance, it known the maximum load transfer capability. So, under the maximum power transfer condition, the voltage source affects by the purely resistive load, and source is not constrained to produce reactive power. But in real-life application customary the transmission system resistance can be negligible compared to the reactance. Therefore, the inductive nature of the system impedance with considering power factor is essential [25].

1.8.2. Maximum Power Transfer Considering Power Factor.

Due to the importance of reactive power consideration in a power system, reference [25] is dealt to find out maximum power based on given power factor ($\cos\theta$). The study is shown, that its value is always negative and refer to the mathematics theory the solution is a maximum. So, power transfer to load maximize when the load impedance becomes equal to the transmission impedance in magnitude, in view of unity power factor. For illustration the Figures 1.12 and 1.13 are given (the simulation results are provided in Annex 1.1). The growing of electricity demand is correlated to the investment on the modern power systems to ensure a reliable system. While, sometimes the economic constraints and electricity market competition criteria do not allow the utility to expand and keep the balance between the demand and supply. Many other factors are involved in voltage instability phenomenon such as Long distance between voltage source and load centers, the insufficient reactive load compensation capability in the system, improper voltage level at voltage source, and etc. which are discussed with details in next chapters.

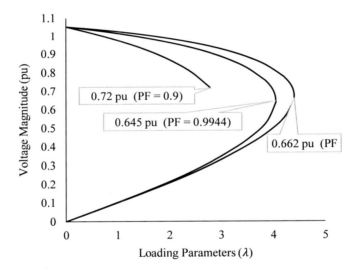

Figure 1.12   The 3rd bus loadability margin at different PF (FGL 3-bus system).

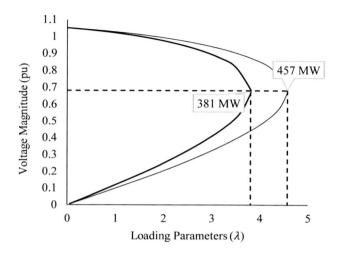

Figure 1.13   Load change (20% increase) at bus 3 (FGL 3-bus system).

1.9.   VOLTAGE INSTABILITY MECHANISM

The increase in electric power demand and limited sources for electric power generation have resulted in an increasingly complex interconnected system, forces the system to operate closer to the limits of stability. The system enters a state of voltage instability when a disturbance, increase in load demand, or change in system condition causes a progressive and uncontrollable decline in voltage. The main factor causing instability is the inability of the power system to meet the demand for reactive power [5]. In general terms, voltage instability may come into being in the form of voltage variation that the power system suffers a progressive fall or rise of voltage of some buses. The obvious possible causes of voltage instability in a power system is greatly loss of load in the area or tripping of transmission lines and so on.

A system enters a stat of voltage instability when a disturbance, increase in load demand, or change in system condition caused a progressive and uncontrollable decline in voltage [26]. Difference reasons could be causing the voltage instability in a power system which the main factors can be listed as following [17]:

- Heavy loaded transmission line.
- Long distance between the voltage source and load centers
- Improper voltage level at voltage source
- The insufficient reactive load compensation capability in the system
- Reactive power imbalance
- Reactive power transfer constraints
- Lack to local reactive power supply (injection)
- System configuration (especially in the radial system)
- Generator limitation
- Transmission line or reactive equipments outages
- Motor stalling

When the voltage instability results in the low voltage in the system that has direct and indirect impact on the system components such as [17]:

A. Power system equipments

When transformer transforms the low voltage to provide accepted level to the secondary, it result in a rise in current flow through the transformer. This high current could lead to thermal overload of the transformer. System load magnitude: There are two type of customer load: motor load and the non-motor load. Motor load is not so much depends on to voltage variation. While the non-motor load that includes constant current and constant impedance types of load. Constant current load directly affected from voltage change, whereas, constant impedance loads (electric heater) varies with the square of the voltage.

B. Customer side

The low voltage impact on the customer side can be considered from reduced light bulb brightness to industrial process interruptions. A generator operates lagging, boosting, pushing or overexcited when supply reactive power and operates vice versa (loading, bucking, pulling or under excited) when absorbing reactive power [17].

C. Power losses

D. Angle stability

Voltage stability versus angle stability is defined that the voltage stability is related to the behavior of customer loads. Angle stability is related to generators magnetic bonds [17].

1.9.1. Voltage Instability from System Planning and Design Criteria Point view

There some factors are associated to lead the system instability namely; environmental pressure, heavy load areas due to increase in electricity consumption, sudden or change in system loading patterns. Such stressed behavior of a power system characterized by slow or sudden change in voltage magnitude (voltage drop), or may sometimes voltage collapse. At the first thought of voltage stability analysis, the stability factors are directly affected from the initial phase of planning and design of a transmission system. These factors are listed as below. Each these component are influenced with voltage stability analysis.

Transmission line (overhead):

A. Parameters

▪   Series Resistance (r, ohm/km)
▪   Series Inductance (l, henry/km)

- Total Impedance ($z = (r + j\omega l)$ ohm/km)
- Shunt Conductance (g, mho/km)
- Shunt Capacitance (c, farad/km)
- Total Shunt Admittance ($y = (g + j\omega c)$ mho/km)

B. The most effective parameters:

- Conductor Size
- Conductor type
- Spacing
- Height above the ground
- Operating frequency
- Thermal rating

C. Classification from Design point of view:

- Short (Can be model by: Positive sequence equivalent circuit)
- Medium (can be model by Lumped parameters model, π or T)
- Long

1.9.2. Voltage Instability from Reactive Power Point view

Obviously, voltage instability or collapse is defined a fall in voltage at first gradual and then rapid due to aggregation of various factors. A system describes unstable in which a small change in power demand leads the system to instability or even collapse. In other word, a system is unstable that is sensitive in load changes. A small increase in power demand at a bus can cause the sudden reaction of unexpected low level voltage, rather than to continue declining in a controlled and predictable manner. The voltage collapse problem is mainly addressed by voltage stability steady state analysis along with the modal analysis methods.

1.9.3. Voltage Collapse

The voltage collapse phenomena may simply refer to improper circulation of reactive power in the system [13]. A stressed power system experiences the system voltage decay lower than acceptable level in which the extremely voltage deviation or uncontrollable steadily voltage reduction in a period of time, can cause a voltage collapse that the system is unable to recover the problem. Complete or a part of the system blackout is often a result of voltage collapse.

Voltage collapse easily can cause the total blackout of the power system. Therefore, voltage collapse prediction has been the focus point of power system planners and operators in order to prevent such unpleasant incident (entire system blackout).

Therefore, power system voltage stability assessment is required a comprehensive investigation mechanism. Because not only it technically affected from many factors such as complexity of the systems, scattering of customers, manageability, and unpredictability behavior of the networks; but also, economic aspects are debatable. So, the sense of techno-economic investigation of power system in order to assure the voltage stability in the system is the primary pillar of power system planning and operation. For obtaining the aforesaid techno-economic criteria, it necessitates to operate closer to the voltage collapse point.

1.10.   CHAPTER 1'S ANNEX

Annex 1.1: FGL's 3-bus system data [19].
Sbase: 100MVA, Vbase: 115kV.

Table A.1.1   FGL's 3-bus system transmission line data.

Branch		S (MVA)	r (pu)	x (pu)
From	To			
1	2	100	0.040	0.0200
1	3	100	0.030	0.0100
2	3	100	0.025	0.0125

Table A.1.2   FGL's 3-bus system generator data.

Bus	Type	Vg	Pg	Qmin	Qmax	R (pu)	X (pu)
Branch-1	2	1.05	2.163	0.0	0.0	0.0100	0.0300
Branch-3	1	1.04	2.000	-0.2	0.4	0.0100	0.0300

Table A.1.3   FGL's 3-bus system load data.

Bus	Type	Load-PL (pu)	Load-QL (pu)
Branch-1	Slack	0.0	0.0
Branch-2	PQ	4.0	2.0
Branch-3	PV	0.0	0.0

Annex 1.2: List of trigonometry identities and functions:

$$\sin 0° = 0 \tag{A.1.2.1}$$

$$\cos 0° = 1 \tag{A.1.2.2}$$

$$\sin(-\theta°) = -\sin \theta° \tag{A.1.2.3}$$

$$\cos(-\theta°) = +\cos \theta° \tag{A.1.2.4}$$

$$\sin^2 \theta + \cos^2 \theta = 1 \tag{A.1.2.5}$$

$$\tan^2 \theta = \sec^2 \theta - 1 \tag{A.1.2.6}$$

$$j^2 = -1 \tag{A.1.2.7}$$

$$j^3 = -j \tag{A.1.2.8}$$

CHAPTER 2

VOLTAGE STABILITY INDICES

The modern power system due to complexity of the various consisting components required timely assessment and monitoring in order to ensure a stable operation. The voltage stability phenomenon is one of the key part of this monitoring and assessment. So, at the last decades, focuses on voltage stability as an essential consideration attracted in power system operation and planning. Hence, the study and research of voltage stability analysis related to estimate the system operation behavior and predict the stability threshold in a power system, have been keen interest of researchers. The most customary methods to estimate the power system the voltage stability margin and approximate the closeness of a power system present operating point to the voltage collapse point have been as following:

- Active and reactive power sensitivity
- Second order performances studies
- Singular value decomposition
- Minimum singular value
- Optimized continuation load flow
- General load flow based analysis and etc.

Among these, the load flow method that includes in the semi-dynamic categories are mostly applied. Especially, a large number of researchers are relied on voltage stability and security studies based on model analysis. All aforesaid categories are associated with merit and demerit, which can be distinguish through performance and applicability of each category. Regardless of the many merits of each method, some demerit of these methods are pointe out in the literature [27, 28, 29] as bellow:

- Intensive computation time
- Huge amount of computation space
- Singularity of the Jacobian matrix
- Complexity in formulation
- Nonlinearity in profile due to change in loading parameters
- Inaccurately predict the collapse point because of its nonlinear behavior when it nears to the collapse point.

The present chapter is dealt with model analysis based voltage stability analysis, in which a voltage stability index is introduced (Danish MSS, Yona A, Senjyu T "A Review of Voltage Stability Assessment Techniques with an Improved Voltage Stability Indicator" (in press, The International Journal of Emerging Electric Power Systems, 2014). In the end, the behavior of the improved indicator performance is outlined. In addition, the improved indicator performance is compared with the indices in the literature. The effeteness of the improved index is tested on IEEE 14-bus, and 30-bus systems as a benchmark to demonstrate the validity of the proposed method. The rest of the chapter is organized as following: in Section 2.1, indices based on the modal analysis and singular value decomposition (SVD) are reviewed. Section 2.2 describes the three voltage stability and security assessment techniques. The preliminary concepts and application of numerical computation methods are defined in Section 2.3, in order to originate the improved indicator. In Section 2.4, a comparative analysis of the indices is evaluated. Finally, the overall obtained results are discussed in Section 2.5.

2.1. LITERATURE REVIEW

With the embrace of literature, this section covers the voltage stability indices, which are based on modal and singular value analysis, as detailed in next subsections.

2.1.1. Modal Analysis-based Indices

In this spirit, the specific framework (voltage stability assessment using modal analysis and numerical techniques) is investigated [16, 30, 31, 32, 33, 34]. The modal analysis method is originally based on the power flow Jacobian matrix. It has applied for purposes of the most critical node identification in the network, which contributes to pursue the system instability. Both the steady state stability analysis and voltage collapse point study can be achieved as arising from a common, well-defined origin [15]. Gao et al. [16] performed the modal analysis technique based on the eigenvalue and associated eigenvectors of the reduced Jacobian matrix of the power flow. In this chapter, the voltage variation versus to the reactive power is considered in respect to the eigenvalues status. Authors pursue the methodology suing an Implicit Inverse Lop-sided Simultaneous Iteration (IILSI) algorithm, in order to calculate the smallest eigenvalues. Berizzi et al. [32] mainly targeted the system security enhancement in respect to the voltage stability. In addition, this chapter figure out voltage collapse point by sensitivity and eigenvalue analysis that was tested on ENEL (the Italian Electric Power Company) transmission system. Barquin et al. [33] investigated the geometry of the reactive power load flow using Jacobian matrix based on maximum power transfer point. As a result, the author assessed distance of voltage stability margin to collapse point. Zambroni de Souza discussed [34] some voltage collapse indices by help of bifurcation technique using center manifold theorem (eigenvalue and singular value characteristics). It has been postulated that the zero eigenvalue is kept constant during the voltage collapse path. The results are relied on characteristics of the tangent vector and reduced Jacobian determinant.

2.1.2. Singular Value Decomposition-based Indices

The other most used technique is singular value decomposition (SVD) method [28, 30, 35, 36, 37, 38]. Indices based on the minimum singular value evaluation are attracted more. Because, these indices can use an estimation tool for reactive power compensation, in which to optimum distribute the resources throughout the system for maximum benefit [36, 39]. The minimum singular value with corresponding to the vectors are used as an indicator for voltage stability and security estimation and monitoring. That estimates, how closely a power system is operating to its collapse point [40]. Sauer and Pai [35] continued with justification of the direct relationship between singularity of load flow Jacobian matrix and the singularity of the system dynamic. Hence, the more focus had been on the role of load flow in a dynamic system. After the comprehensive study of the steady state and the dynamic behaviors of the system; they are concluded that the singularity of load flow Jacobian can decide the maximum loadability. Whereas, it would not observe any instability behaviors associated with dynamic behavior of the system (synchronous machine characteristics or their controls). Lof et al. [36] dealt with an exhaustive study of the SVD technique. His driven index based on load flow Jacobian matrix can proximate the system collapse point and also can recognize the critical buses. Madtharad et al. [30] extracted the static state estimation algorithm by the help of the WLS (Weighted Least Square) algorithm based on SVD method. Calderon-Guizar and Noriega [28] derive an assessment index using the LU-decomposition method. This index is based on start vector for computing the minimum singular value of the load flow Jacobian matrix. This index speeds up the computation of load flow by proposed a new method that reduces the number of iterations. Tiranuchit, and Thomas [37] proposed an index based on minimum singular value, by

41

applying the continuation technique by adding the additional capacitor in the system in order to ensure optimal operation. Hong et al. [38] used load flow equation by employing the non-iterative characteristic of an Incremental Condition Estimation (ICE). This chapter is focused to solve the problem in ICE method when a zero off-diagonal row of the lower triangular matrix encounters.

2.2.   OVERVIEW OF VOLTAGE STABILITY ASSESSMENT TECHNIQUES

For the purpose of voltage stability and security analysis, the dynamic model described by the steady state operation based on the stable equilibrium operating point of the power system are included in these main categories [39]:

- Singular value of the power flow Jacobian matrix
- Power flow Jacobian matrix eigenvalues
- Sensitivity analysis

Whiles, in general the two aspects have been behind the voltage stability analysis examination [16, 39]:

- Proximity of the collapse point.
- Instability mechanism and the key contributing factors.

2.2.1.   Singular Value of the Power Flow Jacobian Matrix Method

The minimum singular value of the power flow Jacobian matrix is used as a measure for the distance between steady state voltage stability limit and supposed operating point. SVD method is one of the basic and most important tool of numerical analysis [41]. The voltage stability mechanism can be enquire from compatible points of view in order to perceive the most variation of the involved components behavior separately. The SVD is a method for identifying and ordering the point's exhibit the most variation [18] and the various relationships among the Jacobian Matrix's correlated variables. The right and left decomposed singular vectors of Jacobian matrix corresponding to the smallest singular value provide information about the sensitivity behavior of critical nodes, and sensitive direction for the changes in active and reactive power injection [42]. The motivation for choice of this assessment index as stated in [28, 30, 35, 36, 37, 38], was that the minimum singular value with corresponding to the vectors are an indicator of voltage stability and security for monitoring how closely a power system is operating to its collapse point [40]. In other word, in voltage stability analysis, the collapse point or singularity of the load flow Jacobian matrix is often attracted as an indicator. The SVD method in keeping with fewer dimensions of the best approximation for the original values can be counted dimensions (data) reduction, and variable ordering method.

However, the SVD method prospered with accurate estimation but also it is time-intensive since it requires a large amount of computational resources for the large matrices [27, 28]. So, there is a need to optimize the computation time in respect to the accuracy tolerance. In this chapter, the pseudo-inverse method is accumulated in order to treat some deficiencies.

2.2.2. Power Flow Jacobian Matrix and Eigenvalue Method

Numbers of voltage stability indices are proposed based on minimum eigenvalues of the Jacobian matrix method [30, 31, 43, 44, 45]. This method, usually, relies on minimum magnitude of the power flow Jacobian eigenvalues in which, the Jacobian matrix becomes zero. The minimum eigenvalues of power flow Jacobian Matrix was used as an indicator. Mostly the indices under this category are proposed for off-line application, which are much time-consuming. These indices are suitable to measure the distance from current operating point to the voltage collapse point [46] or to estimate the voltage stability margin in a power system.

2.2.3. Sensitivity Analysis Method

Those buses, which are easily tending to instability, known weak buses in the system [43]. This behavior often recognizes by bus sensitivity analysis. The sensitivity analysis is used for weak bus identification in literature in order to improve weak buses loadability and enhance the overall stability of the power system [47]. The sensitivity index for bus voltage can be considered the bus voltage variation in respect to $\Delta V_i / \Delta Q_i$ and $\Delta V_i / \Delta P_i$ changes [21, 48]. Sensitivity analysis is effective in weak bus identification. However, sensitivity index alone will not be sufficient to identify weak buses especially in an interconnected system [26]. Beyond the other methods, sensitivity analysis plays important role in prediction of critical nodes in the system. It is important to investigate how this critical point is affected by changing the system conditions [24]. Also, sensitivity analysis can be a useful tool for determining, weak bus, active and reactive power losses, and the reactive power margin (Mvar distant to voltage collapse point) [24].

2.3.   THE IMPROVED INDEX FORMULATION

This section presents an introductory concept of numerical computation methods (SVD, and Pseudo-inverse) as well as the application of these methods in order to integrate the original indices as an improved indicator. Based on the load flow equations, the following customary matrices can be denoted:

$$\begin{bmatrix} \Delta P \\ \Delta Q \end{bmatrix} = \begin{bmatrix} J_{P\delta} & J_{PV} \\ J_{Q\delta} & J_{QV} \end{bmatrix} \begin{bmatrix} \Delta \delta \\ \Delta V \end{bmatrix} = [J] \begin{bmatrix} \Delta \delta \\ \Delta V \end{bmatrix} \qquad (2.1)$$

In static analysis, from the rank of Jacobian matrix $[J]$ point of view, it is assumed that the rank of load flow Jacobian matrix is equal to the rank of system Jacobian (under certain assumptions, such as PV buses are known, loads are constant power (P and Q) types, and the slack node is an infinite bus) [49]. With taking into account the given conditions, the determinant of the load flow Jacobian matrix is identical to the product of all the eigenvalues of the system Jacobian matrix [42]. So, the rank of $[J]$ and reduced Jacobian matrix $[J_R]$ are identical. It implies that the $[J_{P\delta}]$ is nonsingular (it is symmetrical matrix). With pursue the Schur's formula we can obtain (more details are given in [24, 42, 50, 51]):

$$[J_R] = [J_{QV}] - [J_{Q\delta}][J_{P\delta}]^{-1}[J_{PV}] \qquad (2.2)$$

$$det[J] = det[J_{P\delta}] \cdot det[J_R] \qquad (2.3)$$

Every A ($\mathbb{R}^{m \times n}$) matrix, $m \geq n$ can be decomposed as:

$$A = \mathbf{U}\Sigma\mathbf{V}^T \qquad (2.4)$$

$$\Sigma = \mathbf{U}^T A \mathbf{V} \qquad (2.5)$$

$$\Sigma^{-1} = V^T A^{-1} U \tag{2.6}$$

where superscript $(\;)^T$ denotes the transposed matrix. $\mathbf{U}$ $(\mathbb{R}^{m \times n})$, and $\mathbf{V}$ $(\mathbb{R}^{n \times n})$ are termed right and left singular matrixes, which satisfy

$$\mathbf{U}^T \mathbf{U} = \mathbf{V}^T \mathbf{V} = \mathbf{V} \mathbf{V}^T = I_n \tag{2.7}$$

where, $\Sigma = \;< \sigma_1, \ldots, \sigma_n >$ these elements hold on the diagonal of the matrix. Let the σ_i's be called in order, $\sigma_1 \geq \sigma_2 \geq, \ldots, \sigma_n \geq 0$. Which are the square root of the non-negative eigenvalues $\mathbf{A}^T \mathbf{A}$ and are called the singular values of matrix $\mathbf{A}$.

If suppose, $\mathbf{A}$ is nonsingular and one of these conditions satisfy [52]. There is a matrix $\mathbf{A}^{-1}$ such that

- $\mathbf{A}^{-1}\mathbf{A} = \mathbf{A}\mathbf{A}^{-1} = I$
- $\det (\mathbf{A}) \neq 0$
- The eigenvalues of $\mathbf{A}$ are nonzero
- The singular values of $\mathbf{A}$ are nonzero

To infer by linear algebra analogy, the linear equation can be express as $\mathbf{A}x = b$. The solution can be rearranged as $x = \mathbf{A}^{-1}b$; where, we intent to consider the variation of $V - Q$, so change in P is neglected in the formulation ($[\Delta P] = 0$). Therefore, we can obtain

$$\begin{bmatrix} 0 \\ \Delta Q \end{bmatrix} = \begin{bmatrix} J_{P\delta} & J_{PV} \\ J_{Q\delta} & J_{QV} \end{bmatrix} \begin{bmatrix} \Delta \delta \\ \Delta V \end{bmatrix} \Leftrightarrow \begin{bmatrix} \Delta \delta \\ \Delta V \end{bmatrix} = [J]^{-1} \begin{bmatrix} 0 \\ \Delta Q \end{bmatrix} \tag{2.8}$$

$$[\Delta Q] = [J_{Q\delta}][\Delta \delta] + [J_{QV}][\Delta V] \tag{2.9}$$

$$[\Delta \delta] = -[J_{P\delta}]^{-1}[J_{PV}][\Delta V] \tag{2.10}$$

with the substitution the $[\Delta\delta]$ in (9), can be obtained

$$[\Delta Q] = \left([J_{QV}] - [J_{Q\delta}][J_{P\delta}]^{-1}[J_{PV}]\right)[\Delta V] \tag{2.11}$$

$$[\Delta V] = [J_R]^{-1}[\Delta Q] \tag{2.12}$$

$$[J_R]^{-1} = (U\Sigma V^T)^{-1} = \sum_{i=1}^{n} \frac{1}{\sigma_i}[v_i][u_i]^T \tag{2.13}$$

In order to overcome the computational constraints, the Pseudo-inverse method can be introduced as optimization technique for non-square matrix [53]. This method is based on the SVD [51]. For this purpose, a numerical technique (Pseudo-inverse) is employed that is free of computational difficulties [54]. The Pseudo-inverse of matrix $\mathbf{A}$ ($\mathbb{R}^{m\times n}$) is a matrix $\mathbf{A}^+$ ($\mathbb{R}^{n\times m}$) that pursue the following affirmation:

$$\mathbf{A}^+ = (\mathbf{A}^T\mathbf{A})^{-1}\mathbf{A}^T \tag{2.14}$$

Obviously, the $\mathbf{A}^T(\mathbb{R}^{n\times m}) \cdot \mathbf{A}(\mathbb{R}^{m\times n}) = \mathbf{A}^T\mathbf{A}$ matrixes defines a square matrix with these properties:

$$(\mathbf{A}^+)^+ = \mathbf{A} \tag{2.15}$$

$$\mathbf{A}\mathbf{A}^+\mathbf{A} = \mathbf{A} \tag{2.16}$$

$$(\mathbf{A}^T\mathbf{A})^+ = \mathbf{A}^+(\mathbf{A}^T)^+ \tag{2.17}$$

$$(\mathbf{A}^+\mathbf{A})^* = \mathbf{A}^+\mathbf{A} \tag{2.18}$$

where the superscript $(\ )^*$ signifies the conjugate transpose of the matrix. If $m \geq n$ and $\mathbf{A} = \mathbf{U}\Sigma\mathbf{V}^T$, the Pseudo-inverse of $\mathbf{A}$ is

$$\mathbf{A}^+ = \mathbf{V}\Sigma^{-1}\mathbf{U}^T \tag{2.19}$$

If $m < n$, in this case instead of $\mathbf{A}$, we consider the transpose of $\mathbf{A}$ to obtain the Pseudo-inverse.

$$\mathbf{A}^+ = ((\mathbf{A}^T)^+)^T \tag{2.20}$$

There are substitutional expressions for power flow equations in terms of the Pseudo-inverse factorizations of Jacobian matrix. It has been found that, the proposed applied technique can influence the solvability of all probable conditions. From (13), variation of bus voltage related to change in injected reactive power and other involved parameters, can be summarized as voltage stability indicator, as given by

$$[\Delta V] = \sum_{i=1}^{n} \frac{1}{\sigma_i} [v_i][u_i]^T [\Delta Q] \tag{2.21}$$

The above indicator implies that the factorization of the load flow Jacobian, explains that the smallest value of σ can be used as approximate index for describing the voltage stability in the power system. That the corresponding right and left singular vectors indicate sensitive bus voltages and angles, and sensitive direction for the changes in active and reactive power injections. The inverse of the minimum singular value will indicate the greatest change in the state variables. The results were obtained using Neplan® voltage stability function [55] considering these assumptions: the load flow was solved based on Extended Newton-Raphson (ENR) method. The length of the overhead conductors are kept constant (1.0 km).

Note: The vector quantities are denoted as bold or either show in square brackets.

2.4. VOLTAGE STABILITY INDICES COMPARATIVE ANALYSIS

Manifestly, power system is very sensitive to the reactive power deviation. This susceptibility behavior of the power system leads the system to mismatch between generation and load. At long last, it gives rise to frequency deviation in the system. Herein the three techniques are studied, which are mostly based on reactive power deviation in the system. These techniques include the proposed voltage stability indicator (smallest eigenvalue with its associated eigenvectors), bus participation factor, and sensitivity analysis.

In [16, 56] denoted that in view of the voltage collapse mechanism, the bus participation factor can be used for recognition of weakest bus in the system. That the bus participation factors is employed as one of the indicators for identification of the weakest node in the system. Meanwhile, the associated right eigenvectors related to the smallest eigenvalue (the improved indicator) pursues the behavior at the same manner. So, the right eigenvectors of the smallest eigenvalue can be used for weak bus identification as well.

The comparative analysis in Figures 2.1 and 2.2, states the agreement of both techniques, in which the largest participation factor, as well as the largest right eigenvector at the least eigenvalues, indicate the most contribution of the bus to voltage unstable. Although sensitivity analysis is effective for weak bus identification [57, 58] in order to provide sufficient reactive power at the week bus to ensure the secure and reliable operation [59]. From dQ/dV sensitivity point of view; 14, 10, and 7 buses of 14-bus test system, and 26, 30, and 29 buses of 30-bus test system are the 3 top weak buses in the systems respectively. Whereas, from the proposed technique and previous research effort on the same test cases [39, 56], easily can be observed (Figures 2.1 and 2.2) that the ranking of the 3 top weak buses are 14, 10, and 9 buses in 14-bus; and 30, 26, and 29 buses in 30-bus systems. The greatest magnitudes for the indicators denote the weakest bus in the system. The Figure 2.1 demonstrates the S (Sensitivity (% Mvar)), BPF (Bus Participation Factor), and RE (Right Eigenvector) corresponding to the minimum eigenvalue (σ_{min} = 2.079206) of the 14-bus system. The Figure 2.2 indicates the S, BPF, and RE values for 3 top weak buses (30, 26, and 29 buses respectively) corresponding to the minimum eigenvalue (σ_{min} = 0.621337) of the 30-bus system. The recoginaiton of the 3 top weak buses for IEEE 14-bus, and 30-bus test systems (the simulation results are provided in Annex 2.1) , based on sensitivity analysis, bus participation factor and proposed right eigenvector indicators are compared in Table 2.1.

Table 2.1   Stability ranking of the 3 top weak buses through different indicators.

IEEE Test System	Stability Indicator Value			Top 3 Weak Buses Ranking		
	BPF	RE	S	BPF	RE	S
14-bus	0.2374	0.486190	0.2233	14	14	14
	0.2333	0.483920	0.1621	10	10	10
	0.2256	0.476716	0.1417	9	9	7
30-bus	0.174031	0.414541	0.7299	30	30	26
	0.157802	0.399068	0.6733	26	26	30
	0.156804	0.394602	0.6024	29	29	29

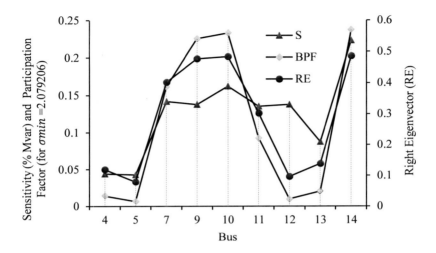

Figure 2.1   The IEEE 14-bus test system's stability indicators namerical values.

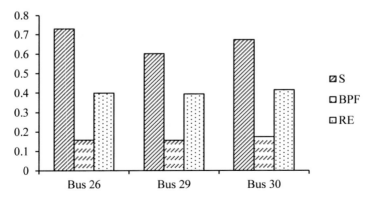

Figure 2.2   The three top, weak buses of the IEEE 30-bus test system's parameters.

2.5.   CHAPTER CONSEQUENCE

There are many other direct techniques beside of the Jacobian matrix method, to diminish the computation time and storage, but tend to fail, if system controls are considered [34]. However, the Jacobian matrix-based techniques have their demerit such: much computation cost and storage requirement, offline application, and lack of perception explicitly. Their merits are also significant from the view of an explicit model, and applicability purpose.

From (13), with considering the solution space and the dimensional subspace of the Jacobian matrix's rank; often, the aforesaid suppositions are not applicable and usually the load flow Jacobian matrix appears that either it is singular or non-square matrix. So in this case, directly the inverse is not exist [60], and under this condition the inverse of Jacobian matrix is known as a juncture for some load flow and voltage stability analysis methods. At the collapse point (saddle-node bifurcation point), the Jacobian matrix is singular at this critical point its eigenvalue is zero [50]. Although, the load flow Jacobian matrix can have a zero eigenvalue whereas the Jacobian matrix is nonsingular [35]. But the static reaction of the system (maximum power transfer point) can motivate the Jacobian matrix to become singular [33].

In order to provide the solution for over-determined system and minimal norm solutions for the under-determined systems according to the rank of A. There is possibility of two general conditions for non-square matrixes, $m \geq n$, and $m < n$. So that, this study investigates both possibility with connive of the previous studies computational restrictions. In order to avoid these constraints, the Pseudo-inverse method was applied as optimization technique for non-square matrix [53], which is based on the SVD [51]. Consequently, through employing the optimization numerical technique the Jacobian matrix would be invertible for all cases regardless the singularity and other constraints.

From (21) can be noted that the greatest right eigenvectors (RE) at the minimum eigenvalue indicates the weakest bus in the system. The magnitude of minimum eigenvalues presents how close the system is to voltage collapse point. If all eigenvalues are positive, then the system is considered to be stable [39]. Inversely, the negative eigenvalues (even if only one) indicates the voltage instability, and zero eigenvalue means that the system in on the border of voltage stability. When $\sigma\_n < 0$, the bus voltage variation in respect to the reactive power are opposite that indicate the unstable voltage [2].

From Section 2.4, can be recalled that the both indicators (the proposed (RE) and BPF) are completely agree on both test systems for the 3 top weak buses identification. However, the S indicator varies on both test systems for three ranking (Table 2.1). Therefore, it could be implied that the sensitivity assessment alone will not be sufficient for the purpose of weak bus identification (often in interconnected system), and the results (that obtain from reactive power deviation, especially near to the collapse point) are not in agreement with aforesaid techniques. The sensitivity analysis could be a useful tool when the system is suffered of heavily load in a stressed situation, in which the dV/dQ and dV/dP sensitivities play and important role in voltage collapse prediction [61].

2.6.   CHAPTER 2'S ANNEX

Annex 2.1: Voltage stability simulation results were obtained by using Neplan® voltage stability function.

Table A.2.1   IEEE 14-bus eigenvalue.

Bus	Mvar (%)
1	64.92836
2	38.533245
3	19.355269
4	18.847389
5	16.098493
6	9.494185
7	7.598687
8	5.387656
9	2.079206

Table A.2.2   Voltage stability indices values for IEEE 14-bus system.

Bus	Sensitivity (S)	At smallest eigenvalue (2.0792)	
		Bus Participation Factor (BPF)	**Right Eigenvector (RE)**
4	0.044	0.0139	0.119854
5	0.0427	0.0064	0.080149
7	0.1417	0.1616	0.401572

9	0.1377	0.2256	0.476716
10	0.1621	0.2333	0.48392
11	0.1353	0.0926	0.302497
12	0.1377	0.0095	0.096489
13	0.0872	0.0198	0.138994
14	0.2233	0.2374	0.48619

Table A.2.3   IEEE 30-bus eigenvalue in descending order.

0.621337
0.971898
1.832569
3.410148
4.469739
5.342615
6.276974
9.447631
13.268052
13.862781
18.766984
21.717075
30.248677
36.290097
43.094696

58.190063
171.658033
322.848264
433.623038
585.414573
1172.003265

Table A.2.4 IEEE 30-bus system Indices' magnitude for three critical buses.

Bus	Sensitivity (S)	At smallest eigenvalue (2.0792)	
		Bus Participation Factor (BPF)	**Right Eigenvector (RE)**
Bus 26	0.7299	0.157802	0.399068
Bus 29	0.6024	0.156804	0.394602
Bus 30	0.6733	0.174031	0.414541

CHAPTER 3

CLASSIFICATION OF VOLTAGE STABILITY

INDICES

Day by day increase in electricity demand and liberalization policy of the electricity markets are persuaded the power systems to operate close to their stability limits. Despite this, the voltage instability can lead the power system to the voltage collapse. A blackout can take place in entire power system or a part of the system due to voltage collapse that can appear abruptly. Instability prediction and continuous monitoring of the power system performance is therefore, known exigent. Both static and dynamic behaviors of a system involve in voltage stability [12].Regardless of the differences, both are correlative to analyze the system stability mechanism. Study of steady state operation is prerequisite to initiate the study of transient stability. Therefore, the steady state equilibrium is a necessary condition for stable transient operation [36, 50, 63].The static voltage stability indices are employed in order to measure the distance from studied current operating point to the voltage collapse point [46]. The static indices can either contribute to identify the critical bus and stability of connected line in the system; as well as, evaluate the stability margin with respect to the system loadability. Power system operation and voltage stability assessment are correlative to each other to ensure reliable, cost-effective operation. In order to know the system performance behavior that how close the system is to voltage collapse, loadability and security limits of the system and overall performance of the system, it is efficacious to employ voltage stability indices.

In the last decades, valuable researches have conducted with the concept of the comparative analysis and classification of voltage stability indices, from different point views. In this context, Massucco *et al.* [64] compared three voltage stability indices with testing in a real power system of the Italian HV transmission grid. Mainly, the authors are focused on functionality of these indices rather than performances behavior. At first, Karbalaei el al. [45] discussed the classification of voltage stability indices, then dealt with the comparison of some indices based on Jacobian matrix and system variables theories. Sinha and Hazarida [43] proposed an index (I_i) based on active and reactive power deviations at the operating point and no-load values. Then compared the effectiveness of the proposed index with three indices through evaluation of these indices by changing the line parameters and load power factor. Reis and Barbosa [56] investigated on some original static voltage stability indices that are previously derived, in order to compare the effectiveness of these indices. Despite theirs promising work, but there is still insufficient information such as an obscure methodology, and covering few indices in the limited extent. Suganyadevi and Babulal [65] performed a valuable comparative analysis of line and nodal indices in which they avoid providing details as much as needed. Huadong Sun el al. [44] published a paper in which they applied the small signal and dynamic analysis to evaluate the accuracy of some indices, which are based on power law model. The authors were found that sometimes the static indices based on load flow model is inexact. Cupelli el at. [66] Investigated four original voltage stability indices, which include various categories based on a different formulation and techniques. This study is performed in view of the indices performance with respect to the load factor change under different operating situations. The RTDS® a (Real-Time Digital Simulator) is used to estimate the real-time behavior of the indices. Finally, authors were found that the voltage collapse point indicators ($VCPI$) has the best performance among the studied indices. As well as, there is other research effort of these author [67] that were compared the suitability of the two indices (Line stability Index (L_{mn}) and $VCPI$) in order to found which one is the most suitable for the purpose of control application. As a result, the simulation indicates that both indices are in agreement related to identifying the weak bus in the system, and generally they pursue the same manner. In [68], the mathematical terminologies and application in examples as well as real systems of some original indices are reviewed. Which is mostly rely on quasi-steady state and dynamic analysis, and are included Voltage Sensitivity Factor (VSF), singular values, eigenvalue decomposition, second order, voltage instability proximity index ($VIPI$), loading margin, direct methods, P and Q angles, test functions and so on. Muhammad Nizam *et al.* [69] compared the power transfer stability index ($PTSI$), which is derived by considering two-bus Thevenin equivalent system, with line index (L index is known as a traditional index for voltage stability) and $VCPI$. In [13], authors are derived a quantitative measure based on operating point of load flow for on-line application. This index is varies in the range of 0 (no load) and 1.0 (voltage collapse point). The index is formulated by using two bus system power flow equations. Then the index is generalized for multi-bus system in view of PQ and PV categories. In [70], the L index is denoted as non-absolute indicator of voltage stability in the system. In 2013, Wang *et al.* [71] extended this index based on alternative generator equivalent model (GEM) instead of ideal constant voltage source. The generalizability of previous research efforts in this context, implies that these efforts were not cover all aspects or have been problematic. Most studies in voltage stability indices comparison and classification have only been carried out in the small context or focused on dynamic analysis.

It is manifest that the adequate picture of voltage stability indices' classification is still ambiguous due to behavioral similarity and intervention of theirs behavior in the system, as well as the variety in application. Nevertheless, this study aims to define comprehensively and compare the performance of original indices, which are proposed and applied globally. On the other hand, distinguish of merit and demerit of the proposed indices are supposed essential, because these indices scale the power system behavior in respect to the system parameters changes, in the form of voltage variation. Despite the literature review of voltage stability indices, the study is concisely reviewed the new voltage stability indices, which are recently proposed [72, 73]. The study can either reveal a unified extensive perspective from various classes (the most outstanding indices from each category) which are recently proposed. The rest of this chapter is organized as follows; in Section 3.1, a review of the voltage stability indices are presented. Section 3.2 describes a broad classification of voltage stability indices with proposing of a novel method. At last, Section 3.3 summarizes the chapter.

3.1. INSIGHT REVIEW OF THE VOLTAGE STABILITY INDICES

Based on the original voltage stability indices as were proposed, the merit and demerit of several indices are discussed from viewpoint of high degree of accuracy in identification of the critical bus and line stability in the power system. At the same time, cursorily has glanced at application, model complexity and perception explicitly of the proposed indices.

3.1.1. Simplified Voltage Stability Index (SVSI)

Pérez-Londoño, *et al.* [72] proposed an improved Simplified Voltage Stability Index (SVSI), to estimate the voltage stability margin in a power system. The index is founded based on Thevenin model and the concept of relative electrical distance (RED) in order to select the nearest generator to a specific load bus, and also the association of electrical variables to improve its performance.

After the identification of the nearest generator to an appropriate load, bus is found with the R_{LG} matrix. The voltage drop on the Thevenin impedance ΔV_i can be estimeted by:

$$\Delta V_i = \sum_{b=1}^{n_j-1} \left| \overrightarrow{V_b} - \overrightarrow{V_{b+1}} \right| \cong \left| \overrightarrow{V_g} - \overrightarrow{V_i} \right| \tag{3.1}$$

R_{LG} is a matrix that indicates the relative locations of load buses with respect to the generator buses.

V_g is the voltage phasor at the nearest generator bus.

V_i is the voltage phasor at the nearest load bus.

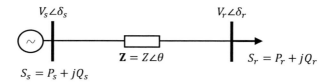

Figure 3.1 Load bus connected to a power system represented by the Thevenin's equivalent [72].

There is observed an addition factor β, due to the involved parameters in the index. Hence, the β is defined as

$$\beta = 1 - \left(max(|V_m| - |V_l|)\right)^2 \qquad (3.2)$$

The proposed correction factor is associated with the highest differences of voltage magnitude between two buses (m and l). Finally the $SVSI_i$ can be represented as:

$$SVSI_i = \frac{\Delta V_i}{\beta V_i} \qquad (3.3)$$

The $SVSI_i$ value close to the unity show voltage unstable bus.

3.1.2. Improved Voltage Stability Index (IVSI)

Chien-Feng Yang, *et al.* [87] proposed and improved voltage stability index for voltage stability enhancement in the power system in which the index formulation is based on power flow variables. Reference to the Kirchhoff's Current Law (KCL) the following equations were expressed as:

$$I_i = V_i \sum_{j=0}^{n} y_{ij} - \sum_{j=1}^{n} y_{ij} V_j \tag{3.4}$$

$$P_i + jQ_i = V_i I_i^* \tag{3.5}$$

$$\frac{P_i + jQ_i}{V_i^*} = V_i \sum_{j=0}^{n} y_{ij} - \sum_{j=1}^{n} y_{ij} V_j \tag{3.6}$$

$$P_i + jQ_i = |V_i|^2 \sum_{j=0}^{n} |y_{ij}| \angle \theta_{ij} - |V_i| \sum_{j=1}^{n} |y_{ij}| |V_j| \angle(\theta_{ij} - \delta_{ij}) \tag{3.7}$$

By distinguish the real and reactive power from the last equation, can be written as:

$$P_i = |V_i|^2 \sum_{j=0}^{n} |y_{ij}| \cos \theta_{ij}$$
$$- |V_i| \sum_{j=1}^{n} |y_{ij}| |V_j| \cos(\theta_{ij} - \delta_{ij}) \tag{3.8}$$

$$P_i = |V_i|^2 \sum_{j=0}^{n} |y_{ij}| \cos \theta_{ij}$$
$$- |V_i| \sum_{j=1}^{n} |y_{ij}| |V_j| (\cos \theta_{ij} \cos \delta_{ij}$$
$$+ \sin \theta_{ij} \sin \delta_{ij}) \tag{3.9}$$

$$Q_i = |V_i| \sum_{j=1}^{n} |y_{ij}| \, |V_j| \sin(\theta_{ij} - \delta_{ij})$$

$$- |V_i|^2 \sum_{j=0}^{n} |y_{ij}| \sin \theta_{ij} \tag{3.10}$$

From combining of the (3.9) and (3.10) can be obtain:

$$\sum_{j=0}^{n} (G_{ij} - B_{ij}) |V_i|^2$$

$$- \sum_{j=1}^{n} |V_j| \, [G_{ij}(\cos \delta_{ij} + \sin \delta_{ij}) \tag{3.11}$$

$$- B_{ij}(\cos \delta_{ij} + \sin \delta_{ij})] |V_j|$$

$$- (P_i + Q_i) = 0$$

From (3.11), since $|V_i|$ is a real number, the following formula needs to be satisfied:

$$SI = \left[\sum_{j=1}^{n} |V_j| \, [G_{ij}(\cos \delta_{ij} + \sin \delta_{ij}) \right.$$

$$\left. - B_{ij}(\cos \delta_{ij} + \sin \delta_{ij})] \right]^2 \tag{3.12}$$

$$+ 4 \sum_{j=0}^{n} (G_{ij} - B_{ij})(P_i + Q_i)$$

$$\geq 0$$

65

Finally, the Improved Voltage Stability Index (IVSI) is proposed as bellow:

$$
IVSI = \left(-4 \sum_{j=0}^{n} (G_{ij} - B_{ij})(P_i + Q_i) \right.
$$

$$
\div \left\{ \left[\sum_{j=1}^{n} |V_j| \left[G_{ij}(\cos \delta_{ij} + \sin \delta_{ij}) \right. \right. \right. \tag{3.13}
$$

$$
\left. \left. \left. - B_{ij}(\cos \delta_{ij} + \sin \delta_{ij}) \right] \right]^2 \right\} \right) \le 1
$$

This index is introduced under bus voltage stability indices category, which is capable to use for both radial and interconnected power systems, as well as it can be used in the same way such other load flow based indices like Voltage Stability Index (VSI) and index L_{mn}. The voltage stability margin is measured between 0 and 1. The index magnitude close to zero indicate a stable system, whereas, as long as the index for any bus in the system is closed to 1.0, indicate the unstable system and voltage collapse may occur (more details are given in [87]). At this chapter, authors are generalized the applicability of the aforesaid bus index for overall system stability evaluation. The aim of the total voltage stability index has been optimizing the setting of compensation devices for N-Bus system. The total voltage stability index is the summation of the N voltage stability indices for all buses of the system. Therefore, the index can be evaluated as bellow:

$$
IVSI_T = \sum_{i=1}^{N} IVSI_i \tag{3.14}
$$

3.1.3. Voltage Deviation Index (VDI)

Chien-Feng Yang, *et al.* [87] are also proposed the other voltage stability index, Voltage Deviation Index (VDI). This index is defined as the absolute value of the deviation of the bus voltage from one per unit, and the generalization for N-bus system is the sum of N voltage deviation indexes separately from each bus, for all the system (more details are given in [87]).

$$VDI_j = \left| 1 - V_j \right| \qquad (3.15)$$

$$VDI_T = \sum_{j=1}^{N} \left| 1 - V_j \right| \qquad (3.16)$$

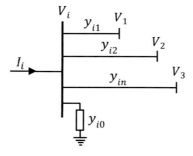

Figure 3.2   A part of the network's single line diagram representation [87].

3.1.4. Voltage Stability Index (VSI)

Haque, M. H, [88] was proposed a simple method using the system variable parameters or local information such bus voltage and current magnitudes in order to determine the distance to voltage collapse from current operating point. The driven index is called Voltage Stability Index (*VSI*). The formulation of this index is relied on power flow (power transfer) equations such a local PQ bus (i) is considerd in a power system as shown in Figure 3.3. From power system concepts, the complex power can be equated in different form as following:

$$\mathbf{S}_i = S_i \angle \theta_i = (P_i + jQ_i) \tag{3.17}$$

$$\mathbf{S}_i = \mathbf{V}_i \mathbf{I}_i^* \tag{3.18}$$

$$\mathbf{V}_i = V_i \angle \delta_i \tag{3.19}$$

$$\mathbf{I}_i = I_i \angle (\delta_i - \theta_i) \tag{3.20}$$

From substitution of the (3.19) and (3.20) in (3.18), the complex power can be obtained. Obviously, with variation in load demand at the PQ bus, the amount of S_i changes as well as V_i and I_i may also change. With applying the Taylor's theorem, author studied, the incremental change between V_i and I_i due to incremental change in S_i as follows. In order to simplify the computation, the higher order terms of the Taylor's theorem expansion are ignored due to their insignificancy.

$$\Delta S_i = \frac{\partial S_i}{\partial I_i} \Delta I_i + \frac{\partial S_i}{\partial V_i} \Delta V_i \tag{3.21}$$

$$\Delta S_i = V_i \Delta I_i + I_i \Delta V_i \tag{3.22}$$

When the load at bus i is increased, the I_i increases, but vise versa the V_i is deacreses. Considering this phenomenon, when the load of a bus approches the critical vlue or the voltage collapse point, the addition load my not increase. So, the rapid reduction in voltage compared to the increase in current due to loadability limit the following results from (3.18). In this case the ΔS_i approachese to zero. Trough tranforming the form of (3.22) by deviding both sides by $V_i \Delta I_i$, can be obtained:

$$V_i \Delta I_i + I_i \Delta V_i \geq 0 \qquad (3.23)$$

$$1 + \left(\frac{I_i}{V_i}\right)\left(\frac{\Delta V_i}{\Delta I_i}\right) \geq 0 \qquad (3.24)$$

The incremental change of ΔV_i and ΔI_i are opposite to each other and both V_i and I_i are positive. Finally, the VSI index is generalized as α (> 1.0):

$$VSI_i = \left[1 + \left(\frac{I_i}{V_i}\right)\left(\frac{\Delta V_i}{\Delta I_i}\right)\right]^{\alpha} \qquad (3.15)$$

At no load, VCI equals unity and at the voltage collapse point its value is zero. The VSI is similar with the Voltage Collapse Indicator in which there is petty change in theirs formulations.

According to the [2], This evaluation of this VSI is very simple and it only requires the magnitude of bus voltage and load current at two different operating points. It is raised to a power of α (> 1) in order to give a more or less linear characteristic to the index. The value of α may depend on the system.

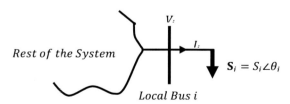

Figure 3.3   Representation of a local bus of a general power system [88].

69

3.1.5. Performance Index (PI)

Mohamed A. and Jasmon G.B. [88] are derived the performance index (PI) that is mainly based on the contingency analysis and load flow techniques. The PI index measures the system stress, and may be used to compare and rank the severity of contingencies. The PI index is derived by substituting the post outage voltage magnitude into the voltage performance index, which is defined as:

$$PI_v = \sum_{i=1}^{N} \frac{w_{vi}}{2n} \left(\frac{|E_i| - |E_i|^{SP}}{\Delta |E_i|^{lim}} \right)^{2n} \qquad (3.26)$$

$|E_i|$ is the voltage magnitude at bus i

$|E_i|^{SP}$ is specified or rated voltage magnitude at busbar i

$\Delta |E_i|^{lim}$ is the voltage deviation limit, above which voltage deviations are unacceptable

n is exponent of PI_v function ($n = 1$ for second order PI_v)

w_{vi} is the real non-negative weighting factor

3.1.6. Voltage Stability Factor (VSF)

Partha Kayal and C.K. Chanda [73] are driven the … index, which is based on power flow analysis. The proposed index is produced from 2-bus system as a part of a distribution system. First the branch current of any branch (i) is supposed as:

$$I_i^2 = \frac{P_{m+1}^2 + Q_{m+1}^2}{V_{m+1}^2} \qquad (3.27)$$

where, P_{m+1}, Q_{m+1}, and V_{m+1} are the active load, reactive load and bus voltage magnitude at but $m + 1$ respectively. The active and reactive power losses in the system can be calculated as follows:

$$Ploss_i = r_i \left(\frac{P_{m+1}^2 + Q_{m+1}^2}{V_{m+1}^2} \right) \tag{3.28}$$

$$Qloss_i = x_i \left(\frac{P_{m+1}^2 + Q_{m+1}^2}{V_{m+1}^2} \right) \tag{3.29}$$

The $Ploss_i$ and $Qloss_i$ are the active and reactive power losses of branch i.

$$I_i^2 = \frac{Ploss_i^2 + Qloss_i^2}{(V_m - V_{m+1})^2} \tag{3.30}$$

From (3.27) can be written:

$$\frac{P_{m+1}^2 + Q_{m+1}^2}{V_{m+1}^2} = \frac{Ploss_i^2 + Qloss_i^2}{(V_m - V_{m+1})^2} \tag{3.31}$$

With substituting the $Ploss_i$ and $Qloss_i$ in (3.31) the equation becomes:

$$(P_{m+1}^2 + Q_{m+1}^2)(r_i^2 + x_i^2) = (V_{m+1}^2 - V_{m+1}V_m)^2 \tag{3.32}$$

Taking the positive root of (3.32):

$$S_{m+1} = \frac{V_{m+1}^2 - V_{m+1}V_m}{Z_i} \tag{3.33}$$

S_{m+1} is the magnitude of complex power at receiving end bus, that for critical power flowing at receiving end can be consider the derivative of (3.33):

$$\frac{dS_{m+1}}{dV_{m+1}} = \frac{V_{m+1}^2 - V_{m+1}V_m}{Z_i} = 0 \tag{3.34}$$

For stable operation of the system the following condition should satisfy:

$$\frac{dS_{m+1}}{dV_{m+1}} > 0 \tag{3.35}$$

$$(2V_{m+1} - V_m) > 0 \tag{3.36}$$

The voltage stability index for any bus $(m + 1)$ can be roduced as:

$$VSF_{m+1} = 2V_{m+1} - V_m \tag{3.37}$$

The voltage collapse threshold for this index is defined zero. When VSF becomes zero the voltage collapse may occur, in this situation the receiving end bus voltage magnitude, becomes half of sending end bus voltage magnitude. A generalization index is proposed as well, as stated in the following:

$$VSF_{total} = \sum_{m=1}^{k-1} (2V_{m+1} - V_m) \tag{3.38}$$

where, k is the total number of buses in the system and V_1 is the magnitude of substation voltage. The higher value of VSF_{total} indicates more voltage stabilt operation.

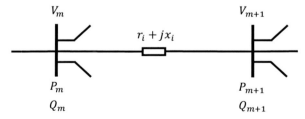

Figure 3.4　Two bus section of radial distribution system [73].

3.1.7. Voltage Collapse Prediction Index (VCPI)

Balamourougan V. *et al.* [61] are proposed the Voltage Collapse Prediction Index (VCPI) in order to predict voltage collapse in power system. This index is derived based on the system variables such as bus voltage magnitude, voltage angle information and system admittance matrix. The advantages of this index is its capability to employ for online application, as well as an online implantation is performed. An N-bus system is considered that the complex power at the kth bus is given by:

$$
S_k^* = |V_k|^2 - (|V_k| \cos \delta_k \\
- j|V_k| \sin \delta_k) \left[\sum_{\substack{m=1 \\ m \neq k}}^{N} (|V_m'| \cos \delta_m' + j|V_m'| \sin \delta_m') \right] Y_{kk} \tag{3.39}
$$

V_m' is given by

$$
V_m' = \frac{Y_{km}}{\sum_{\substack{m=1 \\ m \neq k}}^{N} Y_{kj}} V_m \tag{3.40}
$$

S_k is the complex power at bus k

V_k is the voltage phasor at bus k

V_m is the votage phasor at bus m

δ_k is the voltage angle of bus k

Y_{km} is the admittance between buses k and m

The (3.39) has the similar form such $a - jb$. Hence, the equations for unknowns V_k and δ can be written as

$$f_l(|V_k|, \delta) = |V_k|^2 - \sum_{\substack{m=1 \\ m \neq k}}^{N} |V_m'||V_k| \cos \delta \tag{3.41}$$

$$f_l(|V_k|, \delta) = \sum_{\substack{m=1 \\ m \neq k}}^{N} |V_m'||V_k| \sin \delta \tag{3.42}$$

Solving the two equations for determining the unknowns using the Newton–Raphson technique, a partial derivative matrix is obtained. The determinant of the matrix is equal to zero at voltage collapse resulting in the following equation:

$$\frac{|V_k| \cos \delta}{\sum_{\substack{m=1 \\ m \neq k}}^{N} |V_m'|} = \frac{1}{2} \tag{3.43}$$

Rearranging the (3.43), the VCPI can be obtained at bus k, which is given by:

$$VCPI_{kth\ bus} = 1 - \frac{\sum_{\substack{m=1 \\ m \neq k}}^{N} |V_m'|}{V_k} \tag{3.44}$$

The threshold for VCPI is zero and 1.0 that the index value near to zero state more voltage stable bus. Furthermore, some features of this index are listed as follows:

- Prediction of voltage collapse in the system for every bus.
- The technique is includes the effect of the load changes in the system on the particular bus voltage collapse estimation.

- This technique needs a modest amount of calculations for estimating the VCPI.
- This technique can be used for recombination of the weak bus in the system.
- This technique can be used for both online and offline application.

3.1.8. L index

Jasmon G. B. and Lee L.H.C.C, [80] are presented a method that ranked all contingencies in the system to find out the voltage instable buses as well as voltage collapse cases in the system. For the purpose of the index extraction (L index), first the 2-bus system with its associated parameters are considered. Then tried to derive the equations that characterize the behavior of the single line diagram of the 2-bus system.

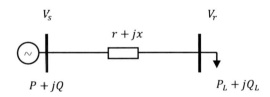

Figure 3.5   Single line diagram of 2-bus system [80].

V_s is the sneding end voltage
V_r is the receiveing end voltage
P is the injection of real power
Q is the injection of reactive power
P_L is real load
Q_L is reactive load
r is resistance of the line
x is reactance of the line

From Figure 3.5, the real and reactive power equations can be derived as:

$$P = \frac{r(P^2 + Q^2)}{V^2 + P_L} \tag{3.45}$$

$$Q = \frac{x(P^2 + Q^2)}{V^2 + Q_L} \tag{3.46}$$

By rearranging the (3.45) and (3.46), can be obtained:

$$x(P - P_L) = r(Q - Q_L) \tag{3.47}$$

With the elimination of Q in (3.45) and rearranging the equation in term of P, the following quadratic equation can be acquired:

$$(r^2 + x^2)P^2 - (2x^2 P_L - 2rx Q_L + r)P \\ + (x^2 P_L^2 + r^2 Q_L^2 - 2rx P_L Q_L + r P_L) = 0 \tag{3.48}$$

Supposing that, the sending end voltage is the reference voltage with a constant magnitude $V_s^2 = 1$. Hence, the (3.48) can be written as:

$$P = \Big\{ [2x^2 P_L - 2rx Q_L + r] \\ - [(2x^2 P_L - 2rx Q_L + r)^2 \\ - 4(r^2 + x^2)(x^2 P_L^2 + r^2 Q_L^2 \\ - 2rx P_L Q_L + r P_L)]^{\frac{1}{2}} \Big\} \div 2(r^2 + x^2) \tag{3.49}$$

Similarly, for reactive power, owing to the symmetry of the equations, can be driven the equation as:

$$Q = \Big\{ [2r^2 Q_L - 2rxP_L + x]$$
$$- [(2x^2 P_L - 2rxQ_L + r)^2$$
$$- 4(r^2 + x^2)(x^2 P_L^2 + r^2 Q_L^2$$
$$- 2rxP_L Q_L + rP_L)]^{\frac{1}{2}} \Big\} \div 2(r^2 + x^2) \quad (3.50)$$

The determinant of the (3.49) and (3.50) quadratic equations can be reduced as:

$$4(xP_L - rQ_L)^2 + xQ_L + rP_L < 1 \quad (3.51)$$

The real and reactive power flows in any line are defined as:

$$P_{i+1} = P_i - r_i \frac{P_i^2 + Q_i^2}{V_i^2 - P_{li+1}} \quad (3.52)$$

$$Q_{i+1} = Q_i - x_i \frac{P_i^2 + Q_i^2}{V_i^2 - Q_{li+1}} \quad (3.53)$$

As well as, the real and reactive loss terms in the (3.52) and (3.53) are given as:

$$R_i = r_i \frac{P_i^2 + Q_i^2}{V_i^2} \quad (3.54)$$

$$X_i = x_i \frac{P_i^2 + Q_i^2}{V_i^2} \quad (3.55)$$

From (3.54), the ratio of real losses between line I and precceding line $i + 1$ can be computed as:

$$\frac{R_{i+1}}{R_i} = \left(r_{i+1} \frac{P_{i+1}^2 + Q_{i+1}^2}{V_{i+1}^2} \right) \div \left(r_i \frac{P_i^2 + Q_i^2}{V_i^2} \right) \tag{3.56}$$

$$\frac{R_{i+1}}{R_i} = \frac{r_{i+1}\left(P_{i+1}^2 + Q_{i+1}^2\right)}{r_i\left(P_i^2 + Q_i^2\right)} \frac{V_i^2}{V_{i+1}^2} \tag{3.57}$$

By relying on Kirchhoff's Current Law (KCL) at not i. The following equation can be obtained:

$$\frac{P_i^2 + Q_i^2}{V_i^2} = \frac{(P_{i+1} + P_L)^2 + (Q_{i+1} + Q_L)^2}{V_{i+1}^2} \tag{3.58}$$

$$\frac{V_i^2}{V_{i+1}^2} = \frac{P_i^2 + Q_i^2}{(P_{i+1} + P_L)^2 + (Q_{i+1} + Q_L)^2} \tag{3.59}$$

For the given distribution network:

$$P = \sum R_i + \sum P_{li} \tag{3.60}$$

$$Q = \sum X_i + \sum Q_{li} \tag{3.61}$$

From (3.59), it can be observed that the losses in the system are ratios of the losses in the first line of the network. Hence:

$$P = r_{eq}(P^2 + Q^2) + \sum P_{li} \tag{3.62}$$

$$Q = x_{eq}(P^2 + Q^2) + \sum Q_{li} \tag{3.63}$$

Since, the r_{eq} and x_{eq} are the equivalent values of resistance and reactnve in the single line respectively. Hence, it can be reduced the real distribution network into a system with only one line. Is make easy to study the voltage collapse for reduced distribution network instead of every line in the network. The roots of $4(xP_L - rQ_L)^2 + xQ_L + rP_L < 1$) for P and Q can be calculated as:

$$L = 4[(xP_L - rQ_L)^2 + xQ_L + rP_L] \tag{3.64}$$

For $L < 1.0$. The generalized form of index is given for the reduced network as bellow:

$$L = 4\left[\left(x_{eg}P_{leg} - r_{eg}Q_{leg}\right)^2 + x_{eg}Q_L + r_{eg}P_{leg}\right] \tag{3.65}$$

$$r_{eg} = \frac{R_{eg}}{(P^2 + Q^2)} \tag{3.66}$$

$$x_{eg} = \frac{X_{eg}}{(P^2 + Q^2)} \tag{3.67}$$

r_{eg} is the equivalent resistance of single line

x_{eg} is the equivalent reactance of single line

P_{leg} is the total real losds in the distribution network

Q_{leg} is the total reacive loads in the distribiution network

When the network is loaded beyond its critical limit, the power becomes imaginary and it is the collapse point.

3.1.9. Power Stability Index (PSI)

Aman M.M. *et al.* [82] derived the PSI for the purpose of the recognition of the most optimum site of Distributed Generation (DG) based on the most critical bus in the system that is sensitive to the variation of the load increase in the system. This index alike many other indices, is driven from simple two-bus system that its single line diagram and associated phasor diagram are presented in Figures 3.6-3.9. From two-bus system can be written as:

$$S_L = P_L + jQ_L = V_r I_r^* \tag{3.68}$$

$$\overline{V_r} = \overline{V_s} - \overline{I_r Z} \tag{3.69}$$

$$I_r = \frac{P_L - jQ_L}{V_r^*} \tag{3.70}$$

From Figures 3.7 and 3.8 can be obtained:

$$I_r = \frac{(P_L - jP_G) - j(Q_L - jQ_G)}{V_r^*} \tag{3.71}$$

With putting the value of I_r in (3.69), the following equations in term of real and imaginary parts can be obtained by:

$$P_L - P_G = \frac{|V_s||V_r|}{V_r^*}\cos(\theta - \delta_s + \delta_r) - \frac{|V_r|^2}{Z}\cos(\theta) \tag{3.72}$$

$$Q_L - Q_G = \frac{|V_s||V_r|}{V_r^*}\sin(\theta - \delta_s + \delta_r) - \frac{|V_r|^2}{Z}\sin(\theta) \tag{3.73}$$

With rearranging, the (3.72) can be written:

$$|V_r|^2 - \frac{|V_s||V_r|\cos(\theta - \delta)}{\cos(\theta)} + \frac{Z(P_L - P_G)}{\cos(\theta)} = 0 \qquad (3.74)$$

$$\delta = \delta_s - \delta_r \qquad (3.75)$$

From (3.74) observes that the quadratic equation for stable voltages should have real roots. There for the determinant of the quadratic equation has to satisfy $b^2 - 4ac > 0$, as a result the proposed index is given as:

$$PSI = \frac{4r_{ij}(P_L - P_G)}{[|V_i|\cos(\theta - \delta)]^2} \leq 1 \qquad (3.76)$$

The voltage stable operation is defined *PSI* less than unity.

V_s is the sneding end voltage
V_r is the receiving end voltage
δ_s is the sending end voltage angle
δ_r is the receiving end voltage angle
P_G is injection of real power
Q_G is injection of reactive power
P_L is real load
Q_L is reactive load
Z is the line impedance
θ is the line impedance angle
r is resistance of the line
x is reactance of the line

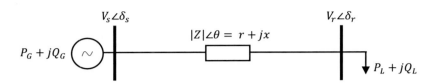

Figure 3.6   Single line diagram of 2-bus system [82].

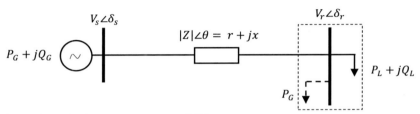

Figure 3.7   Active power support [82].

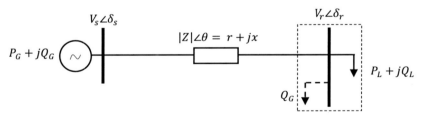

Figure 3.8   Reactive power support [82].

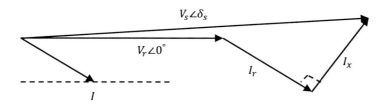

Figure 3.9 Phasor diagram of a simple two-bus system [82].

3.1.10. Stability Index (SI)

Chakravorty, M. and Das, D. [89] proposed a new index for radial distribution networks, which is driven with considering the composite load modeling and power flow analysis. The bellow equation that recalled from [103] is based to obtain the new index.

$$|V(m2)| = 0.707\left[b(jj) + \{b^2(jj) - 4.0c(jj)\}^{1/2}\right]^{1/2} \qquad (3.77)$$

jj	is the branch number
$IS(jj)$	is the sending end node
$IR(jj)$	is the receiving end node
$r(jj)$	is the resistance of branch jj
$x(jj)$	is the reactance of branch jj

From (3.77) can be found that the feasible load flow solution of radial distribution network is exist if

$$b^2(jj) - 4.0c(jj) \geq 0 \qquad (3.78)$$

After some simplifications and substitutions the SI is given by

$$SI(m2) = \{|V(m1)|^4$$
$$- 4.0\{P(m2)x(jj)$$
$$- Q(m2)r(jj)\}\}^2 \tag{3.79}$$
$$- 4.0\{P(m2)r(jj)$$
$$+ Q(m2)x(jj)\}|V(m1)|^2$$

The smallest magnitude of the index at any bus, indicates the most sensitive to the voltage collapse.

3.1.11. Predicting voltage collapse index (V/V_o)

The simple (V/V_o) index [83] is proposed; thus, voltage magnitude (V) is obtained from load flow for the operating point of the system. Where, V_o (no load voltage) is known new values of voltage when the system all loads set to zero. Simply, this index ranking orders the critical buses in the system in respect to the voltage sensitivity. This index indicates an overall picture of the system stability state to determine the weak bus easily. The smallest index value indicates the most sensitive weak bus in the system. This index can be used for on-line and off-line applications. While, with respect to change in loading parameters, this index shows nonlinear profile. As, authors in [2, 67] argued that the (V/V_o) index is poor in all three computational cost, accuracy of collapse point prediction and adequacy to nonlinearity performances.

3.1.12. L Index

Kessel. P and Glavitsch. H, [13] derived L index that is capable to detect the voltage instabilities in the power system, which is included the following features:

- Identification of the vulnerable system states
- Continuously assessment of the quantitative of the real power system state

- Recognition of the critical buses or areas in the system, where countermeasures can be applied
- Predicting of voltage collapse under various contingencies such as loss of generators or lines as well as load variations
- Extension of the concept of security assessment in control centers

The concept of L index is driven based on an analytical analysis of line mode of two-bus system as shows in Figure 3.10.

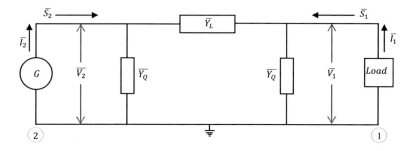

Figure 3.10 Line model of two-bus system [13].

V_1 and V_2 are the nodal voltages at node 1 and 2
I_1 and I_2 are the nodal current at node 1 and 2
S_1 and S_2 are the complex powers
Y_L is series admittance of transmission line π model
Y_G is shunt admittance of transmission line π model

The node 1 can be described in terms of the admittance matrix of the system:

$$\bar{Y}_{11}\bar{V}_1 + \bar{Y}_{12}\bar{V}_2 = \bar{I}_1 = \frac{\bar{S}_1^*}{\bar{V}_1^*} \tag{3.80}$$

$$\bar{S}_1 = \bar{V}_1\bar{I}_1^* \tag{3.81}$$

Equation (3.80) can be brought into the form

$$\bar{V}_1^2 + \bar{V}_0\bar{V}_1^* = \frac{\bar{S}_1^*}{\bar{Y}_{11}} = a_1 + jb_1 \tag{3.82}$$

Above equation consisted from real and imaginary parts, whereby and equivalent voltage $\bar{V}_0$ is substituted

$$\bar{V}_0 = \frac{\bar{Y}_{12}}{\bar{Y}_{11}}\,\bar{V}_2 = -\frac{\bar{Y}_L}{\bar{Y}_L + \bar{Y}_Q}\,\bar{V}_2 \tag{3.83}$$

The analytical solution for $\bar{V}_1$ is given by:

$$V_1 = \sqrt{\frac{V_0^2}{2} + a_1 \pm \sqrt{\frac{V_0^4}{2} + a_1 V_0^2 - b_1^2}} \tag{3.84}$$

$$= \sqrt{\frac{S_1}{Y_{11}}\left(r \pm \sqrt{r^2 - 1}\right)} \tag{3.85}$$

$$r = \frac{V_0^2 Y_{11}}{2S_1} + \cos\left(\phi_{S_1} + \phi_{Y_{11}}\right) \tag{3.86}$$

Whereby ϕ indicates the angle of a complex number. From (3.82) can be written:

$$|\bar{S}_1 - \bar{Y}_{11}^* V_1^2| = V_0 V_1 Y_{11} \tag{3.87}$$

After many calculations and analytical analysis, authors [13] are proposed a stability criterion from above computation as follows:

$$\pm \sqrt{\frac{V_0^4}{2} + a_1 V_0^2 - b_1^2} = 0 \tag{3.88}$$

$$Re\left\{\frac{\bar{V}_1}{\bar{V}_0}\right\} = -0.5 \tag{3.89}$$

Or with the aid of a complex transformation

$$\left|1 + \frac{\bar{V}_0}{\bar{V}_1}\right| = \frac{S_1}{Y_{11}V_1^2} = 1 \tag{3.90}$$

From above equation, an index is proposed, its range is $0 \leq L \leq 1$.

$$L = \left|1 + \frac{\bar{V}_0}{\bar{V}_1}\right| = \left|\frac{\bar{S}_1}{\bar{Y}_{11}^* \bar{V}_1^2}\right| = r + \sqrt{r^2 - 1} \tag{3.91}$$

Finally, after many calculations the generalized L index is given as:

$$L = \frac{MAX}{j \in \alpha_L} \left|1 - \frac{\Sigma_{i \in \alpha_G} \bar{F}_{ji} \bar{V}_i}{\bar{V}_j}\right| \tag{3.92}$$

3.1.13. Equivalent Node Voltage Collapse Index (ENVCI)

Yang Wang, *et al.* [91] introduced equivalent node voltage collapse index (ENVCI) based on equivalent system model (ESM). According to the [91], this index associated with some advantages such affected from both the local network and system outside the local network, real time applications, identification of voltage collapse point when it is near zero. At last, it is demonstrated as prediction and monitoring tool to prevent the system from collapse. In addition, the authors are imputed these futures to the ENVCI with details as follows:

- Accuracy in index modelling, because of the rest of the system outside effect to the local network is considered.
- Distinguish in internal and external impedance consideration that the impedance of the local network are known and there is no need to be estimated using two system states.
- Easiest in calculations and with less computation time consuming; compared to the customary methods based on continuation power flows.
- Fast applicability in real time application with providing the local voltage phasor information, which can be obtained via synchronized phasor measurement units (PMU) or through the state estimation of energy management system (EMS) at control centers of utilities.
- It can be count an emergency remedial action scheme to protect the system from voltage collapse due to its functionality of identification the weak bus at near zero when the system is approaches its voltage collapse point.

The ELNM is formulated based on the outgoing and entering power flow models (more details are given in [91]), as shown in Figures 3.11. After a lot computation, the index is given by:

$$ENVCI = 2(e_k e_n + f_k f_n) - (e_k^2 + e_k^2) \tag{3.93}$$

$\theta_{kn} = \theta_k - \theta_n$ and other parameters are shown in the Figure 3.12.

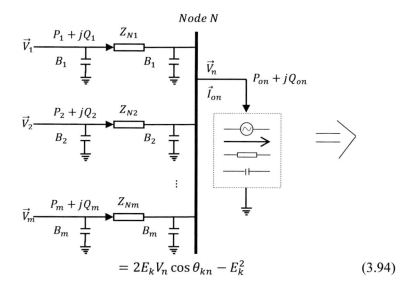

$$= 2E_k V_n \cos\theta_{kn} - E_k^2 \qquad (3.94)$$

Figure 3.11   a) Original local network model [91].

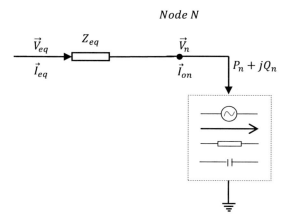

Figure 3.12   b) Equivalent local network model [91].

3.1.14. Line Stability Index (L_{mn})

This index proposed by Moghavvemi, *et al.* [104] that formulated using power transfer concepts in single line power transmission network. The proposed index is given by

$$L_{mn} = \frac{4Qrx}{[|V_s| \sin(\theta - \delta)]^2} \tag{3.95}$$

V_s is the sending end voltage
θ is the line impedance angle
δ is the angle difference between the supply and the receiving end voltages
x is the line reactance
Qr is the reactive power at the receiving end
The value of index must be kept less than 1.0 in order to ensure a stable operation.

3.1.15. Fast Voltage Stability Index (FVSI)

Ismail and Titik Khawa [76] proposed a voltage stability index that symbolized FVSI (Fast Voltage Stability Index). This index is derived based on the voltage collapse occurrence in contingencies conditions the caused by line outage in the power system. The index formulation is similar up to some extend with the concept of voltage stability, in which the discriminant of the roots of voltage or power quadratic equation to be greater than zero. From Figure 3.13 can be obtained

$$I = \frac{V_1 \angle 0 - V_2 \angle \delta}{R + jX} \tag{3.96}$$

The apparent power at bus 2 is

$$S_2 = V_2 I^* \tag{3.97}$$

Rearranging (3.97) yields

$$I = \left(\frac{S_2}{V_2}\right)^* = \frac{P_2 - jQ_2}{V_2 \angle - \delta} \tag{3.98}$$

by separating the real and imaginary parts, and rearranging the above equations yield

$$V_1 V_2 \cos \delta - V_2^2 = RP_2 + XQ_2 \tag{3.99}$$

$$-V_1 V_2 \sin \delta - V_2^2 = XP_2 + RQ_2 \tag{3.100}$$

After some simplification the roots for V_2 will be

$$V_2 = \left\{ \left[\left(\frac{R}{X}\sin \delta + \cos \delta\right)V_1 \right] \right.$$
$$\pm \left[\left[\left(\frac{R}{X}\sin \delta + \cos \delta\right)V_1 \right]^2 \right.$$
$$\left. \left. - 4\left(X + \frac{R^2}{X}\right)Q_2 \right]^{\frac{1}{2}} \right\} \div 2 \tag{3.101}$$

To satisfy the equation with real roots, the discriminant is set greater than or equal to zero. That is given by

$$\frac{4Z^2 Q_2 X}{(V_1)^2 (R \sin \delta + X \cos \delta)^2} \leq 1 \tag{3.102}$$

Since δ is normally very small then,

$$\delta \approx 0 \qquad (3.103)$$

Hence,

$$R \sin \delta \approx 0 \qquad (3.104)$$

$$X \cos \delta \approx X \qquad (3.105)$$

Finally, the FVSI can be defined by

$$FVSI_{ij} = \frac{4Z^2 Q_j}{V_i^2 x} \qquad (3.106)$$

V_1 is voltage at sending bus
V_2 is voltage at receiving bus
P_1 is active power at the sending bus
P_2 is active power at the receiving bus
Q_1 is reactive power at the sending bus
Q_2 is reactive power at the receiving bus
S_1 is apparent power at the sending bus
S_2 is apparent power at the receiving bus
δ_1 is angle of the sending bus
δ_2 is angle of the receiving bus
δ is difference between δ_1 and δ_2
i indicates the sending bus
j indicates the receiving bus
Z is the line impedance
X is the line reactance

Q_j is reactive power at the receiving end

V_i is sending end voltage

In order to ensure a stable power system, the FVSI magnitude must be less than 1.0.

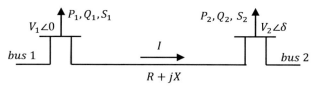

Figure 3.13   2-bus system model [76].

3.1.16. Voltage Collapse Proximity Indicator (VCPI)

Moghavvemi and Farugue [75] proposed a line index for online application that is initiated from maximum power transfer concept. This index is derived from 2-bus system (Figure 3.14) with considering the $\frac{Z_r}{Z_s} = 1$ (the impedance ratio is lower than 1.0 in stable operation) as a voltage collapse predictor. Reference to the impedance ratio ($\frac{Z_r}{Z_s} = 1$), this power transfer boundary can be obtained in term of maximum power transfer as:

$$P_r(\max) = \frac{V_s^2}{Z_s} \frac{\cos \phi}{4 \cos^2 \frac{(\theta - \phi)}{2}} \tag{3.107}$$

$$Q_r(\max) = \frac{V_s^2}{Z_s} \frac{\sin \phi}{4 \cos^2 \frac{(\theta - \phi)}{2}} \tag{3.108}$$

$$P_l(\max) = \frac{V_s^2}{Z_s} \frac{\cos \phi}{4 \cos^2 \dfrac{(\theta - \phi)}{2}} \tag{3.109}$$

$$Q_l(\max) = \frac{V_s^2}{Z_s} \frac{\sin \phi}{4 \cos^2 \dfrac{(\theta - \phi)}{2}} \tag{3.110}$$

V_s is the constanct voltage source
$Z_1 \angle \phi$ is the load impdednace
$Z_s \angle \beta$ is the internal impdednace
$P_r(\max)$ is the maximum transferable real power
$Q_r(\max)$ is the maximum transferable reactive power
$P_l(\max)$ is the maximum real power loss in the line
$Q_l(\max)$ is the maximum reactive power loss in the line

As result, a set of voltage collapse prediction in view of the allowable maximum power transfer limits are given

$$VCPI(1) = \frac{P_r}{P_r(\max)} \tag{3.111}$$

$$VCPI(2) = \frac{Q_r}{Q_r(\max)} \tag{3.112}$$

$$VCPI(3) = \frac{P_l}{P_l(\max)} \tag{3.113}$$

$$VCPI(4) = \frac{Q_l}{Q_l(\max)} \tag{3.114}$$

P_r is the real power transferred to the receiving end

Q_r is the reactive power transferred to the receiving end

P_l is real power loss in the line

Q_l is reactive power loss in the line

The results show some similarity, hence instead of considering four indicator, either real or reactive terms can be considered. In critical situation, both the power and loss indicators are approaches to become 1.0.

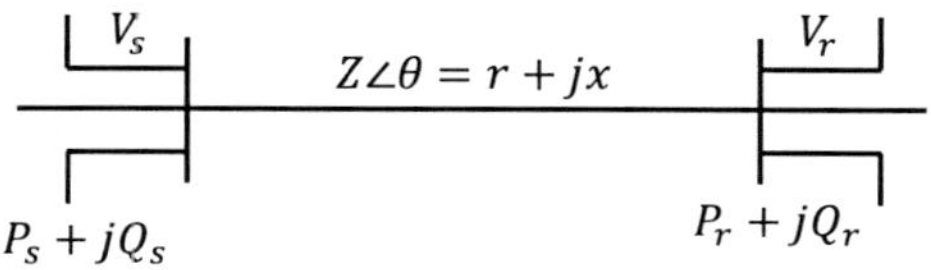

Figure 3.14 Typical transmission line of a power system network [75].

3.1.17. Novel Line Stability Index (NVLS)

Yazdanpanah-goharrizi and Asghari [43] are derived the *NLSI* . Fundamentally, it initiated from load flow equations of 2-bus system and authors claim it effectiveness for point of voltage collapse, weak bus identification and most critical line in an inter-connected system.

$$NLSI_{ij} = \frac{R_{ij}P_j + X_{ij}Q_j}{0.25V_i^2} \qquad (3.115)$$

where, V_i is voltage at sending bus. P_j and Q_j are active and reactive power at receiving end bus, and R_{ij} and X_{ij} are line resistance and reactance between sending end and receiving end bused respectively. The voltage stability tolerance is considered less than 1.0 in order to ensure the line stability.

3.1.18. Line Stability Factor (LQP)

From the line indices that are derived by Mohamed, *et al.* [79], the LQP is formulated based on the power transmission concept in a single line. This index is expressed as:

$$LQP = 4\left(\frac{X}{V_i^2}\right)\left(\frac{X}{V_i^2}P_i^2 + Q_j\right) \tag{3.116}$$

V_i is the sending end voltage
P_i is the sending end real power
Q_i is the receiving end reactive power
X is the line reactance

For maintaining a stable system, the condition LQP must be kept less than 1.0.

3.1.19. Critical Voltage (V_{cr})

Huadong, *et al.* [44] derived a simple index from a single load and infinite bus power system, using load flow equations and eigenvalue theorem.

$$V_{cr} = \frac{E}{\sqrt{2(1 + \cos(\alpha - \phi))}} \tag{3.117}$$

Finally, can be simply expressed as

$$V_{cr} = \frac{E}{2\cos\theta} \tag{3.118}$$

V_{cr} is the critical voltage at receiving end
E is the infinite bus voltage
α is the line impedance angle
ϕ is the power factor angle ($PF = \cos\phi$)
θ is the receiving end voltage angle.

3.1.20. Test Function

References [2, 93] are proposed the test function index based on the quadratic shape of the proposed model. The test function index is more reliable than other Jacobian matrix-based method especially eigenvalue and singular value methods (more details are given in [2, 93]).

$$t_{lk} = \left| e_l^T J J_{lk}^{-1} e_l \right| \tag{3.119}$$

where, J is the Jacobian matrix of the system, e_l is the lth unit vector, i.e., a vector with all entries zero except the lth row, and J_{lk} is defined by

$$J_{lk} = (I - e_l e_l^T)J + e_l e_k^T \tag{3.120}$$

By rearranging the Jacobian matrix with the lth row removed and replaced by row e_l^T. If $l = k = c$, the function shows critical test function as expressed

$$t_{cc} = \left| e_c^T J J_{cc}^{-1} e_c \right| \tag{3.121}$$

It is found that the test function can be used to proximate the voltage collapse in a system, but it will not be able to identify the critical bus, since several buses should be monitored at the same time and that is time-consuming [2].

3.1.21. Second Order Index (i Index)

Berizzi *et al.* [94] proposed a voltage stability index, which is named index i (or second order index). This index is driven based on maximum singular value concept and its derivative. The aim behind this index is to overcome the deficiencies of the previous indices such as minimum singular value index, which are inadequate in non-linearity condition. This index is considered in respect of the system total load and maximum singular value of the inverse Jacobian matrix changes. This index is introduced a useful tool in order to predict the distance to voltage collapse [94]. The range for this index is defined 1.0 at stable condition and zero, when the system tends to collapse.

$$i = \frac{1}{i_0} \frac{\sigma_{max}}{d\sigma_{max}/d\lambda_{total}} \tag{3.122}$$

σ_{max} is the maximum singular value of the Jacobian inverse matrix

λ_{total} is the system total load

i_0 is the value of $\dfrac{\sigma_{max}}{d\sigma_{max}/d\lambda_{total}}$ at the initial operating point.

3.1.22. Tangent Vector Index (TVI_i)

Zambroni de Souza [105] derived the tangent vector index, which is on the load change and corresponding tangent vector concept. This index directly measure the effect of load changes on the vector elements such bus voltage magnitudes and angels. Therefore, it can be counted as good approach to assess how far away the system is operating from the collapse point. This index is given by

$$TVI_i = \left|\frac{dV_i}{d\lambda}\right|^{-1} \tag{3.123}$$

V_i is the voltage at bus i

λ is the load

When the value of the derivative tends to infinity, then $TVI_i \rightarrow 0$.

3.1.23. Power Transfer Stability Index (PTSI)

Muhammad Nizam *et al.* [69] proposed the PTSI, which is derived from 2-bus system based on Thevenin equivalent system. The formulation of this index is given step by step as follows:

$$\bar{I} = \frac{\bar{E}_{Thev}}{\bar{Z}_{Thev} + \bar{Z}_L} \tag{3.124}$$

$$\bar{S}_L = \bar{Z}_L \bar{I} \bar{I}^* \tag{3.125}$$

$$\bar{Z}_L = Z_L \angle \alpha \tag{3.126}$$

$$\bar{Z}_{Thev} = Z_{Thev} \angle \beta \tag{3.127}$$

α is the phase angle of the load impedance
β is the phase angle of the Thevenin impedance

After some substitutions and rearranging can be obtained

$$\bar{S}_L = Z_L \angle \alpha \left| \frac{E_{Thev}}{Z_{Thev} \angle \beta + Z_L \angle \alpha} \right|^2 \tag{3.128}$$

simplification of the above equation yields

$$S_L = \frac{E_{Thev}^2 \, Z_L}{Z_{Thev}^2 + Z_L^2 + 2 \, Z_{Thev} \, Z_L \cos(\beta - \alpha)} \tag{3.129}$$

From differentiating of the apparent power (S_L) with respect to the load impedance (Z_L) the maximum power point can be found as

$$Z_{Thev}^2 - Z_L^2 = 0 \tag{3.130}$$

$$Z_{Thev} = Z_L \tag{3.131}$$

With considering the $Z_{Thev} = Z_L$, the maximum load apparent power is determined by

$$S_{Lmax} = \frac{E_{Thev}^2}{2Z_{Thev}(1 + 2\cos(\beta - \alpha))} \tag{3.132}$$

From power transfer point view, when $Z_{Thev} = Z_L$, the load bus distance to voltage collapse, a power margin is defined as $S_{Lmax} - S_L$. For power margin values equal to zero, it implies that no more power can be transferred. It results that

$$\frac{S_L}{S_{Lmax}} = 1 \tag{3.133}$$

Finally, through rearranging the equations the PTSI index is defined as:

$$PTSI = \frac{2S_L Z_{Thev}(1 + \cos(\beta - \alpha))}{E_{Thev}^2} \tag{3.134}$$

The threshold of the PTSI values are 0 and 1.0, which 1.0 indicates that a voltage collapse has occurred.

3.1.24. Smallest Eigenvalue

Gao *et al.* [16] introduced the model analysis-based indices, which the smallest eigenvalue associated with right eigenvectors, is one of these techniques as expressed:

$$\Delta V = \sum_i \frac{\xi_i \, \eta_i}{\lambda_i} \Delta Q \qquad (3.135)$$

The above equation obviously show the relationship between involved parameters, in which the changes in reactive power, eigenvalue and eigenvectors have directly effect on ΔV.

ΔV indicates the deviation in voltage magnitudes
ΔQ indicates the deviation in injected reactive power
ξ_i is the ith column right eigenvector
η_i is the ith row left eigenvector of Reduced Jacobian matrix
λ is the diagonal eigenvalue matrix of Reduced Jacobian matrix

A system is voltage stable if the eigenvalues of the Jacobian are all positive and an eigenvalue with positive real part indicates that the system is unstable.

3.1.25. Line Voltage Stability Index (LVSI)

Naishan *et al.* [81] proposed the Line Voltage Stability Index (LVSI) that is established the relationship between line reactive power and the sending end voltage. This index is given by

$$LVSI = \frac{4rP_r}{V_s \cos(\theta - \delta)^2} \qquad (3.136)$$

The condition to having a stable system, the value of LVSI must satisfy $LVSI \leq 1.0$.
V_s is the sending end voltage

P_r is the active power at the receiving end

θ is the line impedance angle

δ is the phase

r is the line resistance

3.1.26. Singular Value Indicator

Lof *et al.* [36] proposed a static voltage stability index that is based on a singular value decomposition of power flow Jacobian matrix. The aim of this index is to proximate the voltage instability to collapse and identifies the critical nodes in the system. Supposed that the matrix **A** is and $n \times n$ quadratic (real) matrix.

$$\mathbf{A} = \mathbf{U}\Sigma\mathbf{V}^T = \sum_{i=1}^{n} \sigma_i \, \boldsymbol{u}_i \, \boldsymbol{v}_i^T \qquad (3.137)$$

U and **V** are $n \times n$ orthonormal matrices

$\boldsymbol{v}_i$ and $\boldsymbol{u}_i$ are singular vector and the columns of the **U** and **V** matrices

Σ is a diagonal matrix with

$$\Sigma(\mathbf{A}) = \boldsymbol{diag} \, \{\sigma_i(\mathbf{A})\} \qquad (3.138)$$

$i = 1, 2, \cdots, n$; where $\sigma_i \geq 0$ for all $\boldsymbol{i}$. The order of the diagonal matrix is $\sigma_1 \geq \sigma_2 \geq \cdots \geq \sigma_n \geq 0$. With considering the power flow Jacobian matrix the result is yielded

$$\begin{bmatrix} \Delta\theta \\ \Delta V \end{bmatrix} = \mathbf{V}\Sigma^{-1}\mathbf{U}^T \begin{bmatrix} \Delta F \\ \Delta G \end{bmatrix} \qquad (3.139)$$

From singular value decomposition of the power flow Jacobian matrix these point are observed

A.　The smallest singular value (σ_n) can be used as steady –state stability limit indicator;

B.   The right singular vector $(\boldsymbol{v_n})$ corresponding to smallest singular value $(\boldsymbol{\sigma_n})$ indicated sensitive voltages and angles;

C.   The left singular vector $(\boldsymbol{u_n})$ corresponding to smallest singular value $(\boldsymbol{\sigma_n})$ indicated the most sensitive direction for changes of the active and reactive power injections.

3.1.27. Impedance Ration Indicator

Cheboo *et al.* [12] introduced the impedance ration as a voltage collapse proximity indicator. This index is driven from 2-bus system (Figure 3.15) by employing the Thevenin theorem, as given by

$$\frac{Z_{ii}}{Z_i} \leq 1 \tag{3.140}$$

$$Z_{ii}\angle\beta_i = i^{th} \quad diagonal\ element\ of\ [\mathbf{Z}] \tag{3.141}$$

$$[\mathbf{Z}] = [\mathbf{Y}]^{-1} \tag{3.142}$$

The aim of this indicator is said, the assessment of the validity and robustness of the indicator over the operating range.

$Z_{ii}\angle\beta_i$ is the Thevenin's equivalent impedance
$Z_i\angle\phi_i$ is the impedance of the load

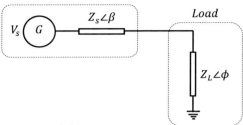

Figure 3.15 2-bus system [1].

In addition, useful information regarding the critical threshold values are formulated in this paper as follows (details are given in [12, 74]):

$$P_{crit} = \frac{V_s^2}{Z_s} \frac{\cos \emptyset}{4 \cos^2 (\beta - \emptyset)/2} \tag{3.143}$$

$$Q_{crit} = \frac{V_s^2}{Z_s} \frac{\sin \emptyset}{4 \cos^2 (\beta - \emptyset)/2} \tag{3.144}$$

$$V_{crit} = \frac{V_s}{2 \cos \frac{(\beta - \emptyset)}{2}} \tag{3.145}$$

where, V_s is constant voltage source (sending end), $Z_s \angle \beta$ is the internal impedance of the network, and $\emptyset$ is the phase angle of the load impedance.

3.2. CLASSIFICATION OF VOLTAGE STABILITY INDICES

Some recent attracted techniques for voltage stability indices have been [74], modal analysis [16], singular value decomposition [36], energy function [84], continues power flow [85], sensitivity analysis methods [35], bifurcations theory [15], minimum eigenvalue [16], and etc. With the embrace of the literature research on voltage stability indices, this study is tried to introduce the voltage stability indices classification for proposed indices as follows:

3.2.1. From Formulation Point view

System variables-based, and Jacobian matrix-based voltage stability indices are often formulated with 2-bus system. Which are based on power flow analysis and the Jacobian matrix [45, 67] and classified into bus and line indices. Jacobian matrix-based *VSI*s can be used to determine the voltage collapse point; in other word, Jacobian matrix-based voltage stability indices demonstrate maximum loadability and determine voltage stability margin. Obviously, due to real-life interconnected power systems complication, the Jacobian matrix computation is much time consuming hence it is not applicable for voltage stability online assessment. This method usually relies on minimum magnitude of power flow Jacobian eigenvalue that via support of linear algebra the Jacobian matrix come up zero.

The system variables-based voltage stability indices deal with power system elements such as weak bus or area voltage assessment, line loadability limit and briefly admittance matrix. The disadvantage is that they are not efficient to appraise roughly voltage stability margin, so they usually apply for online assessment of the crucial element of power system [2]. PMU technology is another category that has been used for power system voltage stability monitoring rather than instability prediction. Nowadays, the PMU hardware technology is known as an accurate and advanced time-synchronized technology for voltage instability monitoring to tracking system dynamics in real-time [86]. The PMU based voltage monitoring techniques are classified into two major classes, which are based on local measurements and relied on Thevenin impedance calculations, and wide-area monitoring (global) measure-ments [86]. However, the Thevenin method has its deficiency due to the parameters variation during the two measurements [72].

3.2.2. From Assessment Point view

Fundamentally, most of the voltage stability indices are distinguished into two categories. Based on the measurement objectives such proximity to voltage collapse point; that predicts, how the system operating close to voltage instability? and voltage instability mechanism identifies the most sensitive and voltage-weak areas [5, 45].

3.2.3. From Application Point view

From application point view, some indices are used to measure proximity to voltage collapse point in off-line and on-line applications [68]. The bus indices constitute a lion's share of this category. Irrespec-tive of some indices, which are formulated based on both Jacobian and system parameter variables, the classification is tend to introduce all the proposed indices into three categories, which are detailed in Table 3.1. The merit and demerit distinguish of the proposed indices are attracted. Because, they play an important role to estimate the power system state in respect to the system parameters changes, in the form of voltage variation. The aforesaid analysis result can abridge in Table 3.2, as a comparative between Jacobian matrix based and system variables based VSIs [66, 67]. There is sufficed to appraise the general characteristics of the Jacobian matrix-based and system variables-based categories, regardless to attend the trivial points related to each index separately. A general detail is described in Table 3.2.

Due to importance of the voltage stability as a prediction and preventing tool in power systems, the indicators of instability phenomenon is become more prominence in power system analysis. Literature survey on voltage stability indices indicates that, due to the lack of an organized, detailed and complete classification of voltage stability indices, there is need for comprehensive study of these indices. Therefore, this section is dealt with detailed illustration of voltage stability indices classification from different point view.

The exact classification of voltage stability indices seems perplexing. From the foundation and performance analysis viewpoints, some proposed indices have typical proximity but vice versa, theirs performance behaviors and accuracies are different. Therefore, A few number of the indices may not exactly fit the classified categories because they are considered from the standpoint of their most characteristics tendency to each category.

Table 3.1 Categorization of voltage stability indices

Type		Index	Abbreviation	Calculation	Stability Threshold	Reference
System parameters (variables)-based	For Bus	Voltage Deviation Index	VDI	$VDI_j = \lvert 1 - V_j \rvert$	Details are given in reference	[87]
		Voltage Collapse Index	VCI	$VCI_i = \left[1 + \left(\dfrac{I_i \Delta V_i}{V_i \Delta I_i}\right)\right]^{\alpha}$	$VCI_i \geq 0$	[64]
		Voltage Stability Factor	VSF	$VSF_{total} = \displaystyle\sum_{m=1}^{k-1}(2V_{m+1} - V_m)$	The greatest magnitude is more stable	[73]
		Performance Index	PI	$PI_v = \displaystyle\sum_{i=1}^{N}\dfrac{w_{vi}}{2n}\left(\dfrac{\lvert E_i \rvert - \lvert E_i \rvert^{SP}}{\Delta \lvert E_i \rvert^{lim}}\right)^{2n}$	Details are given in reference	[88]
		Improved Voltage Stability Index	$IVSI$	$\left[-4\displaystyle\sum_{j=0}^{n}(G_{ij} - B_{ij})(P_i + Q_i)\right]$ $\div \left[\displaystyle\sum_{j=1}^{n}\lvert V_j \rvert\,[G_{ij}(\cos\delta_{ij}\right.$ $+ \sin\delta_{ij})$ $\left. - B_{ij}(\cos\delta_{ij} + \sin\delta_{ij})]\right]^2$	$IVSI \leq 1$	[87]
		Voltage Collapse Prediction Index	$VCPI_{kth\ bus}$	$VCPI_{kth\ bus} = 1 - \dfrac{\sum_{\substack{m=1 \\ m\neq k}}^{N}\lvert V'_m \rvert}{V_k}$	$VCPI_{kth\ bus} < 1$	[61]

Power Stability Index	PSI	$$PSI = \frac{4r_{ij}(P_L - P_G)}{[|V_i|\cos(\theta - \delta)]^2}$$	$PSI \leq 1$	[82]
L Index	L	$$L = \frac{MAX}{j \in \alpha_L} \left| 1 - \frac{\sum_{i \in \alpha_G} \bar{F}_{ji}\bar{V}_i}{\bar{V}_j} \right|$$	$L < 1$	[13]
Stability Index	SI	$$\begin{aligned} SI(m2) &= \{|V(m1)|^4 \\ &- 4.0\{P(m2)x(jj) \\ &- Q(m2)r(jj)\}\}^2 \\ &- 4.0\{P(m2)r(jj) \\ &+ Q(m2)x(jj)\}|V(m1)|^2 \end{aligned}$$	The smallest magnitude is most sensitive to voltage collapses	[89]
Voltage Stability Index	VSI	$$VSI_i = \left[1 + \left(\frac{I_i}{V_i}\right)\left(\frac{\Delta V_i}{\Delta I_i}\right)\right]^\alpha$$	$VSI_i \geq 0$	[90]
Equivalent Node Voltage Collapse Index	ENVCI	$$\begin{aligned} ENVCI &= 2(e_k e_n + f_k f_n) \\ &- (e_k^2 \\ &+ e_k^2) \end{aligned}$$	ENVCI>0	[91]
Sensitivity Analysis	SA	$\Delta V_i / \Delta Q_i$ $\Delta V_i / \Delta P_i$	Details are given in reference	[24]
Novel Line Stability Index	NLSI	$$NLSI_{ij} = \frac{R_{ij}P_j + X_{ij}Q_j}{0.25V_i^2}$$	$NLSI_{ij}$ < 1	[43]

	Fast Voltage Stability Index	$FVSI$	$$FVSI_{ij} = \frac{4Z^2 Q_j}{V_i^2 x}$$	$FVSI_{ij}$ < 1	[76]
	Lmn Index	L_{mn}	$$L_{mn} = \frac{4Qrx}{[\|V_s\| \sin(\theta - \delta)]^2}$$	$L_{mn} < 1$	[77]
	Line Index	L	$$L = 4 \left[\left(x_{eg} P_{leg} - r_{eg} Q_{leg} \right)^2 + x_{eg} Q_L + r_{eg} P_{leg} \right]$$	$L < 1$	[80]
	Voltage Collapse Proximity Indicator	$VCPI$	$$VCPI(1) = \frac{P_r}{P_r(max)}$$ $$VCPI(2) = \frac{Q_r}{Q_r(max)}$$ $$VCPI(3) = \frac{P_l}{P_l(max)}$$ $$VCPI(4) = \frac{Q_l}{Q_l(max)}$$	$VCPI < 1$	[75]
	Line Voltage Factor	LQP	$$LQP = 4 \left(\frac{X}{V_i^2} \right) \left(\frac{X}{V_i^2} P_i^2 + Q_j \right)$$	$LQP < 1$	[79]
	Power Transfer Stability Index	$PTSI$	$$PTSI = \frac{2 S_L Z_{Thev}(1 + \cos(\beta - \alpha))}{E_{Thev}^2}$$	$PTSI < 1$	[69]
	Critical Voltage	V_{cr}	$$V_{cr} = \frac{E}{2 \cos \theta}$$	The critical voltage value	[44]
	Line Voltage Stability Index	$LVSI$	$$LVSI = \frac{4 r P_r}{V_s \cos(\theta - \delta)^2}$$	$LVSI \leq 1$	[81]

For Line

	Tangent Vector Index	TVI	$TVI_i = \left\|\dfrac{dV_i}{d\lambda}\right\|^{-1}$	Depends on load increase	[92]
	Test Function		$t_{cc} = \|e_c^T J J_{cc}^{-1} e_c\|$	Details are given in reference	[93]
		i	$i = \dfrac{1}{i_0}\dfrac{\sigma_{max}}{d\sigma_{max}/d\lambda_{total}}$	$i > 0$	[94]
	Minimum Eigenvalue		$\Delta V = \sum_i \dfrac{\xi_i\,\eta_i}{\lambda_i}\Delta Q$	All eigenvalues should be positive	[16]
	Minimum Singular value		$\begin{bmatrix}\Delta\theta\\\Delta V\end{bmatrix} = V\Sigma^{-1}U^T\begin{bmatrix}\Delta F\\\Delta G\end{bmatrix}$	Details are given in reference	[36]
	Predicting Voltage Collapse		$\dfrac{V}{V_0}$	The smallest index value	[83]
	Impedance Ratio		$\dfrac{Z_{ii}}{Z_i}$	$\dfrac{Z_{ii}}{Z_i} \le 1$	[12]

Jacobian-based

Phasor Measurement Units (PMU)-based	Local Measurement-based	Approximate Approach		$V_{Li} = E_{eq,i} - Z_{eq}I_{Li}$ $Z_{eq} = Z_{LLii}$	Details are given in reference	[95]
		Simplified Voltage Stability Index	$SVSI$	$SVSI_i = \dfrac{\Delta V_i}{\beta V_i}$	$SVSI_i < 1$	[72]
		Voltage Instability Predictor	VIP	$\Delta S = \dfrac{(V_k - Z_{Th}I_k)^2}{4Z_{Th}}$	Details are given in reference	[96]
		Recursive Least Square	RLS	x_k $= x_{k-1} + G_k(y_k - H_k^T x_{k-1})$ G_k $= P_{k-1}H_k(\lambda I$ $+ H_k^T P_{k-1}H_k)^{-1}$ $P_k = \dfrac{1}{\lambda}(I - G_k H_k^T)P_{k-1}$	Details are given in reference	[97]
		Voltage Stability Load Bus Index	$VSLBI$	$VSLBI_k = \dfrac{|V_i(k)|}{|\Delta V_i(k)|}$	Details are given in reference	[98]
	Observa	Sensitivity Rela-ted Eigenvalue		$S_{Qgq} = -g_q^T(g_x^T)^{-1}\Delta_x Q_g$		[86]

Margin Voltage Stability Index	$MVSI$	$VSI = \min$ $\left(\dfrac{P_{margine}}{P_{max}}, \dfrac{Q_{margine}}{Q_{max}}, \dfrac{S_{margine}}{S_{max}}\right)$	Details are given in reference	[99]
Voltage Collapse Proximity Indicator	$VCPI$	$VCPI_{kth\ bus} = \left\| 1 - \dfrac{\sum_{\substack{m=1 \\ m \neq k}}^{N} V'_m}{V_k} \right\|$	$VCPI_{kth}$ <1	[61]

Table 3.2   Comparative analysis between Jacobian matrix-based and system variable-based indices.

Characteristic	Voltage Stability Indices	
	Jacobian Matrix-based	**System variables-based**
Time	More time-consuming	Less time-consuming
Application	Power system voltage stability margin estimation	Power system elements' crucial state recognition (weak bus or stressed area and line identification)
	Measure of the distance from current operating point to the voltage collapse point	Constrains that caused voltage instability phenomenon
Merit	It is very sensitive near the steady state boundary	Response in the overall system load change
	Assess the whole system and could count a centralized measurement	
	Better performance in radial systems than interconnected systems	
	Variety in application in power system such as recognition of the optimum placement of FACTS	

	(flexible AC transmission system) and distributed generator in the system.	
Demerit	Mostly reactive power limits on generators are not considered in during index formulation	Often extracted based on 2-bus system model
	Due to the nonlinearity of the system, this method is not accurate close vicinity of the actual voltage	
	Some indices are based on the computation of path matrix, or RED (related electrical distance) method. Which are computationally expensive.	
	Some indices under this category, shows nonlinear profile due to change in loading parameters	
	Does not accurately predict the collapse point because of its nonlinear behavior when it nears to the collapse point.	

Formulation Concept	Collapse point	Stability margin
	Eigenvalue approach	Maximum power capability
	Stability margin,	Reactive power margin
	PV-PQ Voltage	

Some scholars have argued that the voltage stability indices in use are very different and for convenience they classified these indices in two categories: given state-based, and large deviation-based indices [24, 100]. Despite the existing variety among the indices in view of various aspects, there are two common characteristics among all these classes [2, 84]:

- Proximity of the collapse point.
- Instability mechanism and the key contributing factors.

After all, the proposed classification that is depicted in Figure 3.16, has not been discussed in the literature before. The purpose of Figure 3.16, is to summarize and provide a concise overview of the indices relationship. The aim of the simulations that were carried out on WSCC 9 and IEEE 14, and 30-bus systems (Annex 3.1) [62, 39], was to assess the merit and demerit of the voltage stability indices, which are mostly proposed and applied globally, because of their simplicity in formulation and wide applicability. Further, a brief introduction of the rest indices are sufficed in Table 3.1. For this purpose, a coherent framework is proposed to qualify the merit and demerit of the indices in view of the index performance in order to identify the critical bus and line stability (or proximate the tended buses to collapse) in the power system.

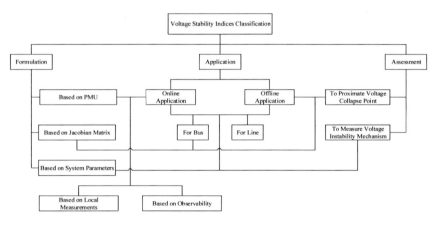

Figure 3.16   Voltage stability indices classification and relationships.

The generalizability of an exhaustive investigation on comparison of voltage stability indices seems problematic due to the multiplicity of indices, and more importantly their variety in theoretical foundations and performances. However, this study from many and various indices in Table 3.1, are studied few numbers. One of the main drawback of the remained indices in Table 3.1, is their high computation cost due to indices complexity and dedicated tools and system parameters requirement. The aim of this study is not to overcome the indices shortcomings. Rather, the idea is to assess the indices theoretical formulation, functionality, applicability, and overall performances. The study is carried out based on this methodology: a) the indices are studied into two categories, node and line indices. b) Such that, each simulated test system in a category is discussed separately with respect to the ranking of the 3 top sensitive critical lines, and 3 top critical buses identification in the system by means of each index. c) Then the overall test systems results are concluded. In order to facilitate in the calculation effort with preserve of the accuracy, the following are assumed: The angles ratio of the V and V_0 in (V/V_o) index calculation are neglected. Because of their close prices and negligible impact on the index magnitude. To avoid ambiguous in the indices calculation, for some cases the magnitude of line parameters such as resistance (R), and reactance (X) are supposed 0.000001, instead of the given zero values. The critical buses ranking by VSF, PSI, and V/V_0 node indices at 30-bus system are chosen by analogy of the other indices. Because of the multiplicity of the indices zero magnitudes of many buses.

The obtained numerical consequences were the result of using power system simulation tools such as PowerWorld® Simulator [101], and MATPOWER a package of MATLAB® [102].

3.3.   CHAPTER CONSEQUENCE

All the indices analyzed in this section depend to two categories, bus and line indices. The critical buses and lines are sorted in descending order in Tables 3.3 and 3.4 that named 1 to 3. Inasmuch as they can withstand a small amount of load before causing voltage collapse.

From Table 3.3, can be noted that the proposed line indices are comparatively agreed on the identification of the top 3 critical branches in the system. However, the WSCC 9-bus systems result indicates that the L index is appeared different in ranking of the 2nd and 3rd critical feeders. While, the obtained result from V_{cr} index is quite diverse with the rest of indices. At 14-bus system, almost all indices are in agreement at first and second ranking order; except, V_{cr} and $LVSI$. Whereas, the indices are varied in the third critical branch identification. Altogether, the 30-bus systems ranking results imply that except the first three indices, all are enormously diverse. So far, a quick conclusion can be drawn that line indices are affected from the system configuration in an interconnected system.

Table 3.3   The obtained three top critical branch ranking by each index.

	Feeder		Voltage Stability Index								
	From	To	NLSI	VCPI		FVSI	L_{mn}	LQP	L	V_{cr}	LVSI
				P	Q						
WSCC 9-bus system	7	5	1	1	1	1	1	1	1	–	1
	9	6	2	2	2	2	2	2	3	–	2
	7	8	3	3	3	3	3	3	2	–	3
	5	4	–	–	–	–	–	–	–	1	–

6	4	–	–	–	–	–	–	–	2	–
8	9	–	–	–	–	–	–	–	3	–
IEEE 14-bus system 4	9	1	1	1	1	1	1	1	1	–
2	3	2	2	2	2	2	2	2	–	1
3	4	–	3	3	–	–	–	–	–	2
12	13	–	–	–	3	3	–	–	–	–
5	6	–	–	–	–	–	3	–	2	–
1	5	–	–	–	–	–	–	–	–	3
4	7	–	–	–	–	–	–	–	3	–
13	14	3	–	–	–	–	–	–	–	–
IEEE 30-bus system 2	5	1	1	1	1	3	1	2	–	1
27	30	2	2	2	–	–	–	3	–	2
29	30	3	3	3	–	–	–	–	3	3
4	12	–	–	–	3	–	2	–	–	–
6	8	–	–	–	–	–	–	1	–	–
6	10	–	–	–	–	1	–	–	1	–
9	10	–	–	–	–	2	–	–	2	–
23	24	–	–	–	2	–	3	–	–	–

Table 3.4, illustrates the response for ranking of the 3 top critical weak buses by each indices. Exclude from VSF and PSI, almost all the proposed bus indices have the same recognition ranking with dissimilarity in

identification of the third critical bus at 14-bus system and first and second weak buses recognition at 30-bus system.

Table 3.4   The obtained weak bus ranking by each index.

IEEE Test system	Bus	Feeder		Index					
		From	To	VSF	PSI	V_j/V_o	BPF	RE	S
14-bus system	14	9	14	–	–	1	1	1	1
		13	14	–	–	–	–	–	–
	10	9	10	–	–	2	2	2	2
	9	4	9	3	3	–	3	3	–
		7	9	–	2	–	–	–	–
	7	4	7	2	1	–	–	–	3
	13	6	13	–	–	3	–	–	–
	4	3	4	1	–	–	–	–	–
30-bus system	26	25	26	2	1	2	3	3	3
	29	27	29	3	2	3	2	2	2
	30	29	30	1	3	1	1	1	1

The comparison shows that virtually the performance of the proposed indices have a high degree of accuracy for assessing the critical node and line that the results are almost close in agreement. By rely on the theoretical formulation of both line and node indices, with considering the simulation results and performances in Tables 3.3 and 3.4. Also (the merit and demerit in Table 3.2), the finding can be noted as follows:

- Almost all indices in a category are in agreement related to identifying the weak buses, and critical lines in the system. Generally, a category indices are pursued the same manner.

- Spite of the indices in the same category have the same theoretical foundation mechanism, still the performance of some indices are in disagreement with the rest indices. For instance, the V_{cr} index in line indices category, and VSF and PSI in node indices category do not draw the same result as other indices. The V_{cr} from formulation point view implies that the model analysis-based indices formulation with respect to the Jacobian matrix singularity assumption is not wholly accurate, especially at the at the collapse point. On the other hand, some studies without considering the generalization, directly applied the quantitative results of customary 2-bus model. While, for complex system with multiple generators and control elements is not adequate solution.

- There are some indexes, which fundamentally are the same, but from driven point of view they are different. In other work, theirs results are the complement of each other in a common concord.

- All indices in a class are coherent to theirs typical theoretical bases, and pursue the same performance. The range of stability for most of the indices are between 0 and 1.0. Someway, it indicates that the indices discernment characteristics of performances are in accord.

- Reis and Barbosa [56] have argued that the line indices can also determine the weakest bus in the system. While, the comparison of the line and node indices in Tables 3.3 and 3.4 have negated this argument. Because, mostly the line indices are driven without taking into account the reactive power generation limits it could cause the misidentification.

- An index with a bad ranking in compliance with other indices, does not imply that the index is useless. Whereas, each indices are functional for a specific application. In the literature is pointed out the demerit of the sensitivity indices that these indices alone will not be sufficient to identify critical node, especially in an interconnected system [26]. However, when the system is suffered of heavily load in a stressed situation, the $\Delta V_i / \Delta Q_i$ and $\Delta V_i / \Delta P_i$ sensitivities indicators play and important role in voltage collapse prediction [61].

- Those indices which are initiated from load flow Jacobian matrix, are not suitable for online application due to theirs prediction insufficiency of voltage collapse. Because of theirs non-linearity properties at the collapse point, as well as high computation requirement [72].

- From the literature is found that for solving the instability phenomenon, there are dynamic factors get involved that cause a high dimensional and multipara meters system [14]. So, it may be wise, to consider the static or semi-dynamic behavior. Most indices that measure stable margin of operating point to voltage instability are based on static analysis using power flow model [44].

From indices classification section can be concluded that partly scholars are argued, and customarily used the terms of the static and dynamic indices in the literature. While, partially for the purpose of voltage stability indices formulation, the dynamic model was considered by the steady state operation based on the stable equilibrium operating point of the power system [2] that from [14], can be naming semi-dynamic analysis.

Therefore, it is arduous to refer to static or dynamic classes. This study is embraced the most used indices in the literature in order to identify the critical bus and line stability in a power system. However, in order to avoid the bulk of the work under the framework of the study, the following aspects of indices are not considered in this paper. The Loadability margin estimation when the system loaded up from the base case to reach the collapse, it experiences different operation behaviors along this path. Assess for different X/R ratios of the transmission system, with keeping the power factor constant; especially for line indices.

3.4.   CHAPTER 3'S ANNEX

Annex 3.1: The line and node indices calculations under different circumstances.

Table A.3.1   The obtained indices' magnitude for critical branch

Branch		NLSI	VCPI		FVSI
From	To		P	Q	
4	1	0	-	-	0
2	7	0	-	-	0
9	3	0	-	-	0
5	4	0	-	-	0
6	4	0	-	-	0
7	5	0.457141	0.736932509	0.736933	0.317457221
9	6	0.327217	0.551864034	0.551864	0.205412355
7	8	0.108727	0.200263408	0.200263	0.078146482
8	9	0			0

Table A.3.1   (continuation)

Branch		LQP	L	Vcr	LVSI	Lmn
From	To					
4	1	0	0	0.512655	0	0
2	7	0	0	0.511373	0	0
9	3	0	0	0.514269	0	0
5	4	0	0	0.499485	0	0
6	4	0	0	0.505745	0	0

7	5	0.305393	0.83900625	0.512356	1.472988201	0.3311052
9	6	0.195565	0.431420866	0.515343	1.2863933	0.2131599
7	8	0.076481	0.45663104	0.513275	0.909300896	0.0799636
8	9	0	0	0.508174	0	0

Table A.3.2   The obtained indices' magnitude for critical branch

Branch		NLSI	VCPI		FVSI	Lmn
From	To		P	Q		
1	2	0.041723	0.05211165	0.052112	0.029621675	0.031626
1	5	0.027323	0.053664748	0.053665	0.013449821	0.0129961
2	3	0.299916	0.603222227	0.603222	0.145540502	0.158298
2	4	0.076555	0.256248432	0.256248	-0.027923844	-0.029982
2	5	0.026044	0.045083448	0.045083	0.0112837	0.0118415
3	4	0.099444	0.277198618	0.277199	-0.030169995	-0.029247
4	5	0.006521	0.011445692	0.011446	0.002863807	0.0028176
4	7	0	0		0	0
4	9	0.356591	0.627703555	0.627704	0.356590282	0.3589291
5	6	0.07274	0.115325447	0.115325	0.072739998	0.0734022
6	11	0.024123	0.028720079	0.02872	0.015360637	0.0155092
6	12	0.040494	0.056754618	0.056755	0.01760098	0.0178617
6	13	0.057598	0.070516198	0.070516	0.03320431	0.0337737
7	8	0	0		0	0
7	9	0.064826	0.114111626	0.114112	0.064825202	0.0648746
9	10	0.027853	0.032929845	0.03293	0.020073907	0.0201163
9	14	0.116444	0.154906644	0.154907	0.059217931	0.0603135
10	11	0.022919	0.028003345	0.028003	0.014804447	0.0147367
12	13	0.148792	0.15512846	0.155128	0.092525866	0.0928123
13	14	0.155423	0.203896609	0.203897	0.078305626	0.0795052

Table A.3.2 (continuation)

Branch		LQP	L	Vcr	LVSI
From	To				
1	2	0.026752	0.527999	0.527999	0.097140945
1	5	0.012704	0.527983	0.527983	0.651156176
2	3	0.137778	0.509657	0.509657	1.250791857
2	4	-0.025188	0.514064	0.514064	0.636275
2	5	0.010191	0.516391	0.516391	0.1137122
3	4	-0.026155	0.496846	0.496846	1.187916261
4	5	0.002602	0.502886	0.502886	0.051265482
4	7	0	0.495065	0.495065	0
4	9	0.35659	0.491635	0.491635	0.000174855
5	6	0.07274	0.494136	0.494136	4.77773E-05
6	11	0.012508	0.517274	0.517274	0.060026116
6	12	0.0143	0.516576	0.516576	0.131357137
6	13	0.026398	0.516382	0.516382	0.14285981
7	8	0	0.516396	0.516396	0
7	9	0.064825	0.512818	0.512818	0.001376514
9	10	0.017582	0.509736	0.509736	0.081531112
9	14	0.048499	0.507436	0.507436	0.346865422
10	11	0.01252	0.508084	0.508084	0.069127813
12	13	0.041648	0.509234	0.509234	0.194360274
13	14	0.063087	0.504769	0.504769	0.447143333

Table A.3.3 The obtained indices' magnitude for critical branch

Branch		NLSI	VCPI		FVSI
From	To		P	Q	
1	2	0.040829	0.050792427	0.050792	0.028895385
1	3	0.010919	0.014938477	0.014938	0.007585635
2	4	0.026048	0.045052501	0.045053	0.011276206
2	5	0.30087	0.604439542	0.60444	0.145826669
2	6	0	0	0	0
3	4	0.006176	0.01041434	0.010414	0.002609098
4	6	0	0	0	0
4	12	0.07499	0.118891841	0.118892	0.074989455
5	7	0.090705	0.115125046	0.115125	0.057375984
6	7	0.058802	0.07838679	0.078387	0.0386867
6	8	0.063398	0.0702055	0.070206	0.053334474
6	9	0	0	0	0
6	10	0.043518	0.108695808	0.108696	0.043517353
6	28	0	0	0	0
8	28	0	0	0	0
9	10	0.007967	0.019898883	0.019899	0.007966678
9	11	0	0	0	0
10	17	0.028633	0.033734145	0.033734	0.02059127
10	20	0.012902	0.01767585	0.017676	0.00643366
10	21	0.053035	0.060349918	0.06035	0.037360687
10	22	0	0	0	0
12	13	0	0		0
12	14	0.041984	0.059037913	0.059038	0.018051019

12	15	0.031106	0.041296468	0.041296	0.014679599
12	16	0.024647	0.02937138	0.029371	0.015701355
14	15	0.084992	0.09163833	0.091638	0.040839478
15	18	0.020048	0.027428989	0.027429	0.009061192
15	23	0.023879	0.028382608	0.028383	0.014939354
16	17	0.058128	0.072986174	0.072986	0.043887473
18	19	0.039604	0.051041895	0.051042	0.020694215
19	20	0.004651	0.006144183	0.006144	0.002260905
21	22	0	0		0
22	24	0.0823	0.086120355	0.08612	0.06338884
23	24	0.112158	0.121864537	0.121865	0.085002814
24	25	0	0		0
25	26	0.068102	0.072452792	0.072453	0.048854314
25	27	0	0		0
28	27	0	0		0
27	29	0.034381	0.042956211	0.042956	0.018252083
27	30	0.173159	0.252141728	0.252142	0.056013018
29	30	0.135086	0.197026775	0.197027	0.043749241

Table A.3.3 (continuation)

Branch		Lmn	LQP	L	Vcr	LVSI
From	To					
1	2	0.0310744	0.025997	0.098337212	0.52767	0.090772386
1	3	0.0083082	0.007057	0.012527058	0.52543	0.025650757
2	4	0.0118532	0.01018	0.028476133	0.51566	0.112409942
2	5	0.160899	0.138008	0.328598596	0.50665	1.144735735
2	6	0	0	0	0.5128	0
3	4	0.002668	0.002327	0.007126896	0.50382	0.03010354
4	6	0	0	0	0.49661	0
4	12	0.0757245	0.074989	0.085385582	0.48891	4.50637E-05
5	7	0.0563883	0.049579	0.119787331	0.49235	0.340584256
6	7	0.0395283	0.034978	0.09272165	0.49284	0.206965597
6	8	0.0537404	0.049309	0.39519504	0.49482	0.170861069
6	9	0	0	0	0.49027	0
6	10	9.9869167	0.043517	0.045080286	0.06397	2.27972E-07
6	28	0	0	0	0.49504	0
8	28	0	0	0	0.49455	0
9	10	0.5637133	0.007967	0.00954225	0.0665	2.13041E-07
9	11	0	0	0	0.50967	0
10	17	0.0737685	0.017952	0.041426624	0.50263	0.01411637
10	20	0.0065182	0.005359	0.014111878	0.50097	0.04240821
10	21	0.0376328	0.030728	0.097034106	0.50193	0.121570373
10	22	0	0	0	0.50197	0
12	13	0	0	0	0.51065	0
12	14	0.0183289	0.014659	0.046906472	0.50847	0.136465701

12	15	0.0149431	0.011672	0.035572384	0.50824	0.088769537
12	16	0.0158558	0.012805	0.028024412	0.50924	0.061544138
14	15	0.0409835	0.018357	0.092755535	0.50151	0.120696655
15	18	0.0091584	0.007301	0.021616528	0.49755	0.062840994
15	23	0.0150415	0.011999	0.026091741	0.49812	0.058727501
16	17	0.0440292	0.040854	0.070101281	0.50263	0.239539698
18	19	0.0207567	0.016627	0.043741276	0.49231	0.115489427
19	20	0.0022532	0.001809	0.005017176	0.49186	0.014409058
21	22	0	0	0	0.4962	0
22	24	0.0639164	0.044869	0.098570945	0.49575	0.125606794
23	24	0.0852595	0.068605	0.12586848	0.4924	0.222962198
24	25	0	0	0	0.49107	0
25	26	0.0493389	0.033735	0.07095236	0.4881	0.108666623
25	27	0	0	0	0.49042	0
28	27	0	0	0	0.48512	0
27	29	0.0186823	0.014258	0.036055342	0.49025	0.084977094
27	30	0.0583534	0.043683	0.189629084	0.48792	0.515014545
29	30	0.0444823	0.034177	0.139543964	0.47839	0.435559311

Table A.3.4 The obtained indices' magnitude for weak bus identification

| Bus | Branch | | VSF | PSI | Vj/Vo |
	From	To			
4	2	4	1.080872	1.93368064	0.94908
	3	4	1.004157	5.533106601	0.94908
5	1	5	1.107938	0.263729008	0.954419
	2	5	1.074834	0.306090238	0.954419
	4	5	1.016568	0.244039621	0.954419
7	4	7	0.976958	0	0.949303
9	4	9	0.986545	3.55711E-05	0.929802
	7	9	1.067908	1.96145E-05	0.929802
10	9	10	1.060878	1.045255773	0.925681
11	6	11	1.083193	0.302194356	0.93142
	10	11	1.045101	0.532718621	0.93142
12	6	12	1.085044	0.665506183	0.930355
13	6	13	1.089898	0.645425171	0.926024
	12	13	1.060002	0.36546909	0.926024
14	9	14	1.076698	2.150384317	0.912321
	13	14	1.065465	2.537073855	0.912321

Table A.3.4   (continuation)

Bus	Branch		BPF	RE	S
	From	**To**			
4	2	4	0.0139	0.119854	0.044
	3	4			
5	1	5	0.0064	0.080149	0.0427
	2	5			
	4	5			
7	4	7	0.1616	0.401572	0.1417
9	4	9	0.2256	0.476716	0.1377
	7	9			
10	9	10	0.2333	0.48392	0.1621
11	6	11	0.93142	0.0926	0.48392
	10	11			
12	6	12	0.0095	0.096489	0.1377
13	6	13	0.0198	0.138994	0.0872
	12	13			
14	9	14	0.2374	0.48619	0.2233
	13	14			

Table A.3.5 The obtained indices' magnitude for three weakest buses

Bus	From	To	VSF	PSI	Vj/Vo
26	25	26	1.036053	0.369924625	0.93985
29	27	29	1.044453	0.439171371	0.945386
30	29	30	1.016233	2.707961113	0.934087

Table A.3.5   (continuation)

Bus	From	To	BPF	RE	S
26	25	26	0.174031	0.414541	0.7299
29	27	29	0.157802	0.399068	0.6733
30	29	30	0.156804	0.394602	0.6024

CHAPTER 4

VOLTAGE STABILITY RESTORATION

In order to improve the overall power system reliability and ensure the system voltage stability, as well as to reduce power diversity when suffers from unexpected contingencies such as generation outages, tripping of a transmission line, or sudden increase in load demand. There is a need to resort an effective method. These methods are detailed in the next section.

4.1. METHODS OF IMPROVING VOLTAGE STABILITY

Here, some methods are listed in order to improve the voltage stability in power system.

1. Local reactive power injection using passive shunt compensators

Compensators are divided into two class: Passive and active compensators:

Passive: shunt reactors and capacitors, and series capacitors
Active: synchronous capacitors and thyristor controlled capacitors and

135

reactors

2. Line-length Compensation:

Therefore, it increase in power transfer capability and hence improve voltage stability margin. Inductive reactance of the line is reduced which is equivalent to reduction in line length. This method is known as line-length compensation.

3. Load Shedding

4. Set up additional lines

5. Using FACTS devices

Customary the first two option are technically versatile and economically cost-effective. However, recently research efforts reveal that FACT is got more attractive. Load shedding is the least common preferred option [18].

4.1.1.   Enhancement of Voltage Stability by Compensations

In recent years, voltage stability considerations have been recognized as an essential part of the power system planning and operation. For reactive power re-compensation, three factors are almost essential, and these factors are:

- Recognition of the proper bus i.e. the sensitive weakest but in the system,
- Appropriate placement of re-compensation device to be effective to all network,
- Determination of the amount of reactive power to be injected to improve the overall voltage stability of entire power system and maintain the stability.

Load in a power system consists mostly in two forms, inductive and capacitive types that store reactive power there in magnetic fields and electric fields respectively. That the return cycles of the stored reactive power to the system are opposite one another [18]. In other word, capacitive load inject the existing reactive power in the system, whiles, the inductive load observe the reactive power from system. Power system is very sensitive to the real and reactive power deviations. This susceptibility behavior of the power system leads the system to mismatch between generation and load. At long last, it gives rise to frequency deviation in the system [18].

From the planning point of view, planners try to keep balance the reactive power in the system with the generators. Because the much amount of reactive power causes the voltage drop in the system due to strong current. Eventually, the suitable preventive action to prevent unbalance reactive power circulation in the system is to install capacitor banks near the inductive load or where (the bus) is sensitive to the reactive power change.

4.1.2. Compensator Devices

for reactive power re-compensation, three factors are almost essential and these factors are recognition of the proper bus i.e. the sensitive weakest but in the system, appropriate placement of re-compensation device to be effective to all network, and determination of the amount of reactive power to be injected to improve the overall voltage stability of entire power system and maintain the stability.

4.1.3. Local Reactive Power Injection

In general, in this method, the passive devices such as shunt capacitor bank and synchronous condenser are using. According [24], the reactive power in a system as a static var compensation should be generated close to the consumption point in order to minimize the reactive power transfer, reference to the following reasons:

- To reduce the gradient of voltage magnitude during high active power transfer.
- To reduce system losses related active and reactive power.

- To prevent from damaging temporary overvoltage following load rejections.
- To minimize the size of power system equipments such as lines, transformers, etc.

From various effective aspect on voltage stability in power system, generator plays a significant role for providing reactive power up to its field and armature currents limitations [22].

4.2.   SHUNT COMPENSATION (REACTIVE POWER)

In recent years, stability considerations have been recognized as an essential part of the power system planning and operation. The system enters a state of voltage instability when a disturbance, increase in load demand, or change in system condition causes a progressive and uncontrollable decline in voltage. The main factor causing instability is the inability of the power system to meet the demand for reactive power [5]. Due to the tendency of increasing reactive power demand of the system because of a combination of events and system conditions; additional reactive power demand may lead to voltage collapse, causing a major breakdown of part or all of the system. It is extremely important to ensure the stability of the system in such situations. The reactive power come to be voltage dependent, when the generator reactive power supply capability is affected from field current limit [106].

Beside of the ability of the power system to maintain voltage magnitude at buses in acceptable level, the power transfer in view of various loading scenarios and voltage control are the significant indicators to rank a power system, voltage stable. These indicators generally refer to the power system operators, in a voltage stable power system, operators have control on both voltage magnitudes and power transfer [18]. Most of the existing proposed approaches have focused on voltage stability based on model analysis technique. These methods predominantly use Jacobian sensitivities for improving voltage stability margin [30, 31, 107, 108]. In this chapter, a quick method for determining the bus optimum loadability limit by using shunt capacitors to prevent the risk of voltage instability is presented. The method identifies the weak bus in order to enhance its stability and improve the bus load ability for future expansion and demand. This method is tested on modified IEEE 14-bus system.

4.2.1. Shunt Capacitors

It is a common practice in power system to install capacitors in weak buses to improve voltage stability, power factor, bus loadability and hence to reduce power and energy losses. Voltage stability has been called load stability [111]. The use of shunt capacitors has been increased because of economic (low cost) and technical (easy and quick installation and can be utilized virtually anywhere in power system) [107], benefits associated with the installation of capacitors are influenced in power system.

The extent of the benefits from capacitor banks installation depends on electrical network configuration and its load states. The net profit achieved is the amount saved by reducing losses after discounting the investment in equipment acquisition and its installation [108].

4.2.2. Weak Bus Identification

In an electric power system, heavy loading bus may lead to voltage instabilities or collapses or in the extreme to complete blackouts. From the index proposed in [109] the weakest bus identification is obtained by Extended Newton-Raphson Power Flow (ENRFP). Bus 8 operating point is shown in Figure 4.1, and it is determined the base-case in this study. Most probably bus 8 is the weakest bus.

The power flow assessment results are shown in Table 4.1. On this table, bus

8 is highlighted as weakest bus with 2.93% overload of its maximum loadability margin (106%) with significant variation in voltage. The gradient vectors ΔP and ΔQ and it has been proved that $0° < \beta < 180°$ in the upper part of the V-P curve, while $0° > \beta > -180°$ in the lower half. Where, β is the angle between the gradient vectors ΔP and ΔQ. At the maximum, the gradient vectors are aligned and thus β is equal to $0°$ or $\pm 180°$ [109]. As expected, bus 8 is in a critical situation: its load is 282.31 MW, whereas, the normal operation limit is between 243.46 MW ~ 274.54 MW. The angle β is $153°$ it is at the upper part of P-V curve, and the maximum is $180°$. As a result, bus 8 is the one of the weakest bus to have optimized. The present study has been effective for those buses in the interconnected network that endure of lack of reactive power generation limit or the system excessive reactive power demand.

Algorithm

The step by step procedure of the present algorithm is given as follows:

Step 1: Carry out load flow by Extended Newton-Raphson Method.

Step 2: Identify the weakest bus based on weakest bus index.

Step 3: Check β. If satisfies, we have to install shunt capacitor at that bus with maximum feasible value of shunt capacitor with iteration method in order to optimum loadability and stability.

Step 4: Check the weak bus parameters and go to step 1.

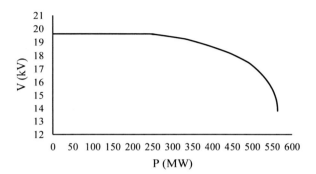

Figure 4.1   The weakest bus P-V curve.

Table 4.1 The enhanced algorithm results.

Bus Load (%)

Bus Number	Before Shunt Capacitor Placement	After Shunt Capacitor Placement
1	106	106
2	104.5	104.5
3	101	101
4	102.38	100.9
5	102.8	101.76
6	107	105.93
7	105.29	99.22
8	108.93	96.47
9	103.57	99.14
10	103.41	99.55
11	104.81	102.31
12	105.38	104.04
13	104.73	103.14
14	102.25	99

Different stability margin are considered for power system (5% or 6%) [107]. in this study minimum stability threshold is set to 6%.

4.2.3. Case Study

The study has been conducted on modified IEEE 14-bus system. Weak bus index is computed by ENRPF and results are tabulated (Table 4.1) which illustrates the results of load buses and compares particularly the weak bus situation before and after placement of shunt capacitor.

Figure 4.3 represents the significant changes in bus loadability, voltage stability, power factor and losses. Bus 8 has been overloaded with 2.93 % from its maximum available loadability margin and after 40 Mvar injection, in addition which, comes over to, overload, it has been increased 3.5 % extra in the loadability margin. It means that even in the most severe condition the bus is capable supply at 3.5% load increase.

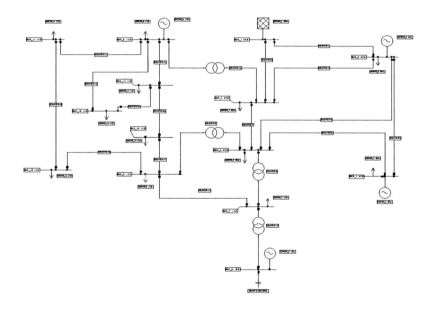

Figure 4.2   Modified 14-bus test system. The bus 8 is the target bus.

It is considered the V-P characteristics with constant load power factor. Voltage stability in fact, depends on how variation in Q, as well as P in the load area, affects the voltages at the load buses. Often, more useful characteristic for certain aspects of voltage stability analysis is the Q-V relationship, which shows the sensitivity and variability of bus voltages with respect to reactive power injections or absorptions [5].

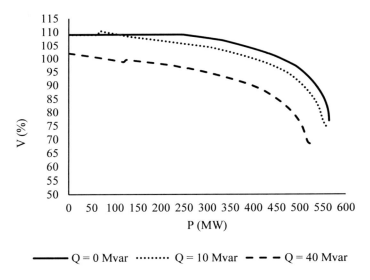

Figure 4.3 P-V curve before and after shunt capacitor placement.

From the sensitivity point of view, there is also change in sensitivity of load buses that the comparison results are shown in Figure 4.4. The relation between required capacitive compensation Q and the expected increased real power demand P_{Di} for a given reactive power demand Q_{Di} may be computed from generic equation as [110]:

$$Q = \propto (D_{Di})P_{Di} + \beta(Q_{Di})$$ (4.1)

This equation is the approximate straight line representation between required capacitive compensation for an expected rear power demand at the given reactive power demand. The constants α and β are the functions of the reactive power demand Q_{Di}[110].

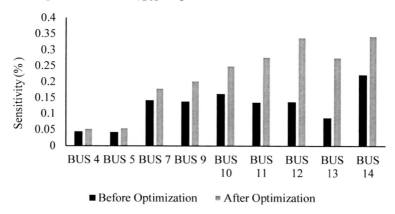

Figure 4.4 Load buses representation before and after optimization.

From the point of view of voltage instability, power system voltage instability may be initiated by a disturbance or an accident under the conditions that are usually characterized by shortage of reactive power reserves [31]. Hence, voltage stability of power system has been closely linked with the reactive power reserves of the system [31]. On the other hand, power factor and voltage stability are correlative. Therefore, power factor correction is counted an essential task, with shunt capacitor place in the system might be able to ensure overall power quality. Figure 4.5 is shown the different condition of system respect to the variation in shunt capacitor MVars. At the initial state system power factor is 0.879, through iteration for Q = 10 MVar, Q = 20 MVar, Q = 30 MVar and Q = 40 MVar the power factor is changed to 0.961, 0.998, 0.992, 0.958 respectively, and after its maximum value gradually decrease to 0.958. Stability of the system and the sign of sensitivities are changed abruptly at a loadability limit [112]. The slower forms of voltage instability are often analyzes as steady-stat problems; power flow simulation is the primary study method. Snapshots in time following an outage or during load buildup are simulated. Besides this post-disturbance power flows, which other power flow based methods are widely used: P-V curves and V-Q curves. Conventional power flow programs can be used for an approximate analysis [21].

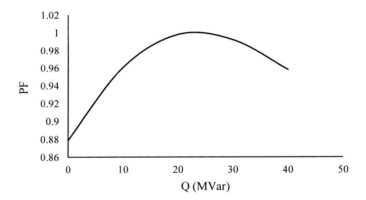

Figure 4.5 Power factor variations vs. reactive power change.

CHAPTER 5

LOAD SHEDDING

The increase in power demand and limited sources for electric power have resulted in an increasingly complex interconnected system, forced the system to operate closer to the limits of stability. Voltage stability is considered to be one of the keen interest of industry and research sectors around the world since the power system is being operated closed to the limit whereas the network expansion is restricted due to many reasons such as lack of investment or serious concerns on environmental problems [6]. A system enters a stat of voltage instability when a disturbance, increase in load demand, or change in system condition caused a progressive and uncontrollable decline in voltage [5]. Voltage stability depends on how variations in reactive, as well as active power in the load area, affect the voltages at the load buses. Frequently, in the literature voltage stability indices and load shedding has been focused through researches. Reference [57], introduced multi criteria integrated voltage stability index was proposed for weak buses identification with application of P-V and P-Q curves. The relationship of reactive power reserves and VSM (Voltage Stability Margin) is quantitatively analyzed in [110]. Sinha [43] has proposed comparative study of voltage stability using index-L and modal analysis and summed with three indices up. Reactive power compensation method using shunt capacitors and reactive power planning are conducted in [113, 114]. Modal analysis and Jacobian matrix are used to evaluate VSM in [30, 31]. Whereas, in [107], introduced a method with using shunt capacitors to improve VSM and applied the modal analysis methodology as well. In [115] presented a method for determining the location and quantity of the load to be shed through mathematical calculation of voltage stability indicator. L Index, study of voltage collapse index are bade based on voltage collapse point theory, which in this point, power flow becomes unsolvable [43, 115]. The others indices are also counted as voltage stability limits, such as sensitivity index and singular value index [116], load proximity index [117, 118], and line stability index [119]. These indexes have various consideration during there's foundation.

In this chapter predominantly is used Jacobian eigenvalue and eigen-vectors indexes via support of linear algebra with ranking of sensitivity for assessment index of voltage stability and improving VSM. The present method is leant on load shedding at appropriate worse bus/area. The test systems show that this assessment index method can be obtained to distinguish the weak and sensitive buses in the power system for the sake of load shedding; simultaneously, the criterion voltages magnitude can be acquire on weak buses through load shedding. At different conditions, critical and normal operation the case studies are pursued to indicate the accuracy and effectiveness of the proposed assessment index.

5.1.   APPLICATION OF LOAD SHEDDING

The study has been conducted on test cases with IEEE 14-bus and 23-bus systems. The practical application of the voltage stability assessment index (14) derived in Chapter 2. For the purpose of analysis, Jacobian matrix, eigenvalue, eigenvectors, and sensitivity are calculated by voltage stability function of NEPLAN® software. The simulation results are provided in Appendix 4.1 to 4.4. The voltage stability and load shedding were analyzed for two cases. Case 1 present validity of assessment index with analysis of weak bus and bus sensitivity. Case 2 presents load shedding and VSM improvement through numerical representation. Today different stability margins are considered for power system 5% or 6%. In this study minimum stability margin is set 5% [107].

5.1.1. Case 1: Weak Bus Recognition and Bus Sensitivity

The IEEE14-bus system is used in this case. The14, 10 and 9 buses were recognized as weak buses for analysis which is depicted in Figure 5.1. Sensitivity analysis is effective in weak bus identification, however, sensitivity index alone will not be sufficient to identify weak buses especially in an interconnected system.

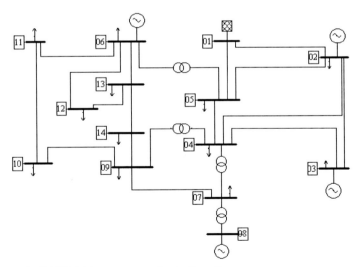

Figure 5.1 IEEE 14-bus system single line diagram.

To obtain an accurate result from the improved assessment index in Chapter 2, the least eigenvalue and greatest right eigenvector should be considered. The 14, 10 and 9 buses having the greatest right eigenvectors at the minimum eigenvalue. Therefore, these buses were known the weakest buses in the system. Meanwhile, bus 7 is the fourth weakest bus in the system while its sensitivity is greater than the sensitivity of bus 9. Table 5.1 demonstrates the indexes with the minimum eigenvalue 2.079206 and maximum eigenvectors 0.48619, 0.48392, 0.476716 corresponding to bus 14, 10 and 9 respectively.

5.1.2. Case 2: Load Shedding Performance Evaluation and Voltage Stability

In this case, 23 bus system (Figure 5.2) is simulated using load shedding method to enhance voltage stability and change the eigenvalue to its maximum possible value and eigenvectors to minimum values. In this case, bus 26 (weakest bus) was selected for load shedding. At the first scenario, 6% load and in the second scenario completely connected load (55.758 MVA) are shed at bus 26. The bus voltage is changed from stressed condition 0.948 pu to 0.95 pu and 0.985 pu at bus 26 respectively. These scenarios represent bus 26 transition to a stable region via increasing in eigenvalue and decreasing eigenvectors, Table 5.2.

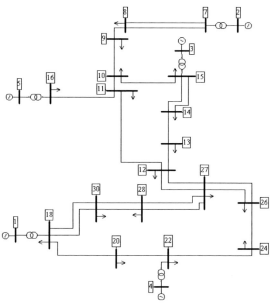

Figure 5.2   Single line diagram of 23-bus system.

This test system is selected from NEPLAN® [55], the voltage stability of a 220 kV transmission network with five power stations is analyzed. The module contains 23 buses.

Table 5.1   Load shedding performance evaluation for enhancement voltage stability.

At $\sigma_{min} = 2.079206$		
Bus	v_i	$\Delta V/\Delta Q$
14	0.486190	0.223312
10	0.483920	0.162144
9	0.476716	0.137696
7	0.401572	0.141678
11	0.302497	0.135261
13	0.138994	0.087230
4	0.119854	0.043989
12	0.096489	0.137652
5	0.080149	0.042725

The results obtained in Table 5.2, it can be seen that by using load shedding matching with the principle of assessment index. At the first scenario, the eigenvalue is changed from 2.076186 to 2.086527 and at the second scenario the eigenvalue increase from 2.086527 to 2.21654 and also there is declined in eigenvectors, as well. Bus 26 can be identified as the weakest bus with the highest voltage collapse point, the lowest reactive power margin and the highest percentage change in voltage. All these can be evidence of assessment index validity and the load shedding effectiveness. Voltage stability steady-state analyses can be assessed by obtaining voltage profiles or shortly P-V curves of critical buses as a function of their loading conditions [120]. Figure 6.3 shows an increase of voltage magnitude at the operating point, likewise improvement of voltage stability of the entire system with 47 MW load decrease at bus 26 as a function of the parameter value.

152

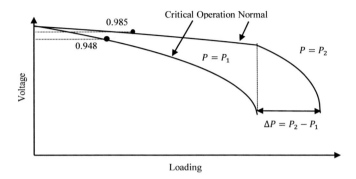

Figure 5.3   The aim of load shedding at bus 26.

Table 5.2 The 23-bus calculation result with $\Delta\sigma_{min} = 0.1403541$

Bus	Critical operation		Normal operation	
	$\sigma_{min} = 2.076183$		$\sigma_{min} = 2.216540$	
	v_i	$\Delta V/\Delta Q$	v_i	$\Delta V/\Delta Q$
26	0.417374	0.107948	0.408093	0.099490
27	0.411640	0.096384	0.406467	0.089750
12	0.393176	0.091369	0.390434	0.085775
28	0.377982	0.134475	0.379584	0.128950
13	0.365707	0.096258	0.365063	0.091385
24	0.262055	0.116076	0.262258	0.112248
11	0.222030	0.090070	0.227594	0.087888
10	0.181515	0.098856	0.191070	0.097655
9	0.173424	0.162855	0.186474	0.161465
30	0.157115	0.072784	0.160910	0.071747
14	0.122397	0.050822	0.124804	0.050089
8	0.065302	0.074600	0.070781	0.074373
15	0.009206	0.003155	0.009551	0.005132
22	0.007082	0.004064	0.007237	0.004061
20	0.004291	0.090098	0.004374	0.090063
16	0.001144	0.001117	0.001174	0.001117
18	0.000316	0.000150	0.000327	0.000150
7	0.000017	0.000020	0.000019	0.000020

5.2.   CHAPTER 5'S ANNEX

Annex 4.1: The 23-bus system simulation results under different states of load shedding (the load change in weak bus).

Table A.4.1   The 23-bus system's V-Q self-sensitivities.

V-Q Sensitivity (%/Mvar)		After load shedded	
Bus	Initial state	6% load shedded	Completely load shedded
9	0.162855	0.162732	0.161465
28	0.134475	0.134066	0.12895
24	0.116076	0.115802	0.112248
26	0.107948	0.107316	0.09949
10	0.098856	0.098738	0.097655
27	0.096384	0.095883	0.091385
13	0.096258	0.095878	0.090063
12	0.091369	0.090934	0.08975
20	0.090098	0.090096	0.087888
11	0.09007	0.089855	0.085775
8	0.0746	0.074581	0.074373
30	0.072784	0.072709	0.071747
14	0.050822	0.050767	0.050089
22	0.004064	0.004064	0.004061
15	0.003155	0.003155	0.003152
16	0.001117	0.001117	0.001117
18	0.00015	0.00015	0.00015
7	0.00002	0.00002	0.00002

Table A.4.2 The 23-bus system's eigenvalues

Eigenvalue (Mvar / %)		
Initial state	**6% load shedded**	**Completely load shedded**
2.076186	2.086527	2.21654
4.410785	4.415275	4.465852
9.98735	10.002469	10.206046
10.937777	10.948464	11.039589

Table A.4.3 The 23-bus system's bus participation factors.

Eigenvalue (Mvar / %)	Bus	Factor (-)
2.076186	26	0.174038
	27	0.17038
	12	0.156043
	28	0.142829
	13	0.135354
	24	0.067046
	11	0.050003
	10	0.03261
	9	0.027922
	30	0.024267
	14	0.015357
	8	0.003993
4.410785	9	0.543693
	10	0.193548

	8	0.10735
	11	0.06272
	28	0.024293
	26	0.017576
	27	0.014742
	24	0.012141
	12	0.008845
	13	0.008641
	30	0.005615
9.98735	28	0.38657
	24	0.288921
	30	0.247987
	26	0.016149
	11	0.014478
	13	0.011469
	9	0.008777
	12	0.007203
	8	0.005137
	10	0.004688
10.937777	24	0.280569
	20	0.22402
	11	0.172135
	10	0.08234
	9	0.080303
	8	0.061132

	13	0.028939
	28	0.021828
	30	0.019053
	12	0.016272
	14	0.010284

Table A.4.4 The 23-bus system's branch participation factors

Eigenvalue (Mvar / %)	Branch	Factor (-)
2.076186	54	1
	16	0.653866
	17	0.653866
	56	0.613691
	57	0.613691
	11	0.585295
	13	0.576305
	8	0.464219
	20	0.455116
	21	0.455116
	7	0.45506
	4	0.387082
	55	0.351603
	2	0.252556
	3	0.252556
	5	0.230508
	10	0.16472
	9	0.16472
	14	0.142306
	15	0.142306
	26	0.132624
	25	0.132624
	12	0.11414

	6	0.082224
	22	0.036804
	15	0.022395
	53	0.01052
4.410785	4	1
	5	0.948417
	2	0.757696
	3	0.757696
	8	0.672069
	6	0.346854
	7	0.303772
	54	0.237168
	56	0.147581
	57	0.147581
	16	0.128747
	17	0.128747
	20	0.121252
	21	0.121252
	55	0.085234
	11	0.069457
	14	0.032979
	15	0.032979
	25	0.029285
	26	0.029285
	10	0.01707

	9	0.01707
	12	0.011158
9.98735	54	1
	20	0.749995
	21	0.749995
	56	0.688418
	57	0.688418
	55	0.360715
	16	0.299038
	17	0.299038
	5	0.244038
	2	0.167308
	3	0.167308
	4	0.138965
	7	0.102477
	8	0.087054
	11	0.086871
	13	0.079841
	25	0.071078
	26	0.071078
	22	0.045741
	9	0.040347
	10	0.040347
	52	0.027917
	53	0.015489

	6	0.012787
10.937777	54	1
	5	0.755013
	2	0.534314
	3	0.534314
	4	0.400376
	55	0.376509
	8	0.341548
	7	0.306563
	53	0.235221
	20	0.229336
	21	0.229336
	56	0.200711
	57	0.200711
	52	0.171448
	13	0.134725
	11	0.118358
	16	0.069545
	17	0.069545
	22	0.064924
	10	0.064316
	9	0.064316
	25	0.043737
	26	0.043737
	14	0.030403

	15	0.030403
	6	0.014448
	15	0.011146

Table A.4.5 The 23-bus system's generator participation factors

Eigenvalue (Mvar / %)	Generator	Factor (-)
2.076186	GEN3	1
	GEN1	0.71508
	GEN4	0.589835
	GEN5	0.352995
	GEN2	0.293007
4.410785	GEN2	1
	GEN3	0.360488
	GEN5	0.263102
	GEN1	0.217985
	GEN4	0.160646
9.98735	GEN1	1
	GEN4	0.62025
	GEN3	0.200052
	GEN2	0.164258
	GEN5	0.085386
10.937777	GEN4	1
	GEN1	0.805074
	GEN2	0.595862
	GEN3	0.478319
	GEN5	0.304187

Table A.4.6 The 23-bus system's right eigenvectors.

Eigenvalue (Mvar / %)	Bus	Right Eigenvector (-)
2.076186	8	0.065302
	7	0.000017
	9	0.173424
	10	0.181515
	11	0.22203
	16	0.001144
	15	0.009206
	14	0.122397
	13	0.365807
	12	0.393176
	27	0.41164
	28	0.377982
	30	0.157115
	18	0.000316
	26	0.417374
	20	0.004291
	22	0.007082
	24	0.262055
4.410785	8	0.330215
	7	0.000088
	9	0.746587
	10	0.431692
	11	0.242633

	16	0.001254
	15	0.004712
	14	-0.027122
	13	-0.090258
	12	-0.091447
	27	-0.118225
	28	-0.152052
	30	-0.073531
	18	-0.000149
	26	-0.129458
	20	-0.002414
	22	-0.002987
	24	-0.108587
9.98735	8	0.073634
	7	0.00002
	9	0.096704
	10	-0.068401
	11	-0.11896
	16	-0.000619
	15	-0.004101
	14	-0.056019
	13	-0.105728
	12	-0.083868
	27	-0.050089
	28	0.618321

	30	0.498755
	18	0.00094
	26	-0.126391
	20	-0.049487
	22	-0.015778
	24	-0.541082
10.937777	8	0.254896
	7	0.000068
	9	0.293527
	10	-0.287725
	11	-0.411667
	16	-0.002143
	15	-0.009732
	14	-0.099444
	13	-0.16776
	12	-0.125883
	27	-0.0434
	28	0.146645
	30	0.137371
	18	0.000728
	26	0.021543
	20	0.464144
	22	0.024465
	24	0.532808

Annex 4.2: The 23-bus system simulation results, after the weak bus 6% load shed.

In this case the system parameters are recalculated based on the new values of the eigenvalues due to 6% load mitigate at the critical bus. The simulation results are given in the following tables.

Table A.4.7   The 23-bus system's bus participation factors after 6% load shed of the critical bus.

Eigenvalue (Mvar / %)	Bus	Factor (-)
2.086527	26	0.173658
	27	0.170175
	12	0.155932
	28	0.142935
	13	0.135327
	24	0.067054
	11	0.05009
	10	0.032774
	9	0.028153
	30	0.024332
	14	0.01538
	8	0.004031
4.415275	9	0.543706
	10	0.193262
	8	0.107433
	11	0.062436
	28	0.024405

	26	0.017635
	27	0.014817
	24	0.012188
	12	0.008914
	13	0.008709
	30	0.005649
10.002469	28	0.386003
	24	0.288669
	30	0.248387
	26	0.016119
	11	0.014609
	13	0.011501
	9	0.008827
	12	0.007223
	8	0.005187
	10	0.004757
10.948464	24	0.27079
	20	0.248413
	11	0.166876
	10	0.080041
	9	0.07773
	8	0.059359
	13	0.028033
	28	0.021294
	30	0.018656

	12	0.015773
	14	0.009975

Table A.4.8 The 23-bus system's Branch Participation Factors after 6% load shed of the critical bus.

Eigenvalue (Mvar / %)	Branch	Factor (-)
2.086527	54	1
	16	0.666595
	17	0.666595
	57	0.619642
	56	0.619642
	11	0.576392
	13	0.574728
	8	0.469612
	20	0.459009
	21	0.459009
	7	0.448893
	4	0.394466
	55	0.348681
	3	0.258031
	2	0.258031
	5	0.236327
	9	0.162699
	10	0.162699
	14	0.143309
	15	0.143309
	26	0.134615
	25	0.134615

	12	0.114147
	6	0.082716
	22	0.03695
	15	0.021783
	53	0.010677
4.415275	4	1
	5	0.954107
	2	0.759036
	3	0.759036
	8	0.666445
	6	0.352324
	7	0.295389
	54	0.233936
	57	0.147216
	56	0.147216
	16	0.130116
	17	0.130116
	20	0.120572
	21	0.120572
	55	0.085027
	11	0.067261
	14	0.032411
	15	0.032411
	26	0.029027
	25	0.029027

	9	0.016436
	10	0.016436
	12	0.011194
10.002469	54	1
	20	0.758041
	21	0.758041
	56	0.702583
	57	0.702583
	55	0.383349
	16	0.314833
	17	0.314833
	5	0.251518
	2	0.171519
	3	0.171519
	4	0.142314
	7	0.103093
	8	0.088697
	11	0.08641
	13	0.081217
	26	0.069793
	25	0.069793
	22	0.046091
	10	0.040058
	9	0.040058
	52	0.028157

	53	0.016365
	6	0.014039
10.948464	54	1
	5	0.775476
	2	0.545763
	3	0.545763
	4	0.408834
	55	0.400422
	8	0.345997
	7	0.308828
	53	0.257479
	20	0.233554
	21	0.233554
	56	0.206308
	57	0.206308
	52	0.188202
	13	0.140881
	11	0.118163
	16	0.07379
	17	0.07379
	22	0.067148
	9	0.063893
	10	0.063893
	25	0.042618
	26	0.042618

	14	0.029503
	15	0.029503
	6	0.012677
	15	0.010881

Table A.4.9 The 23-bus system's right eigenvectors after 6% load shed of the critical bus.

Eigenvalue (Mvar / %)	Bus	Right Eigenvector (-)
2.086527	8	0.06564
	7	0.000018
	9	0.17421
	10	0.182049
	11	0.222269
	16	0.001146
	15	0.009229
	14	0.122557
	13	0.365771
	12	0.392994
	27	0.411316
	28	0.378161
	30	0.157411
	18	0.000316
	26	0.416787
	20	0.004298
	22	0.007095
	24	0.262133
4.415275	8	0.330396
	7	0.000089
	9	0.746623
	10	0.431373

	11	0.242039
	16	0.001251
	15	0.004704
	14	-0.02728
	13	-0.090569
	12	-0.091747
	27	-0.118453
	28	-0.152367
	30	-0.07379
	18	-0.000149
	26	-0.129586
	20	-0.002425
	22	-0.002996
	24	-0.108809
10.002469	8	0.074017
	7	0.00002
	9	0.096986
	10	-0.068913
	11	-0.119474
	16	-0.000621
	15	-0.00412
	14	-0.056182
	13	-0.105857
	12	-0.083964
	27	-0.050135

	28	0.617753
	30	0.49935
	18	0.000941
	26	-0.126198
	20	-0.05023
	22	-0.015807
	24	-0.540814
10.948464	8	0.251305
	7	0.000067
	9	0.288853
	10	-0.28378
	11	-0.405322
	16	-0.00211
	15	-0.009603
	14	-0.098032
	13	-0.165185
	12	-0.123986
	27	-0.042879
	28	0.144909
	30	0.136114
	18	0.00075
	26	0.020834
	20	0.489439
	22	0.024762
	24	0.523702

Annex 4.3: The 23-bus system simulation results, after the weak bus completely load shed. In this case the system parameters are recalculated based on the new values of the eigenvalues due to completely load shed at the critical bus. The simulation results are given in the following tables.

Table A.4.10   The 23-bus system's bus participation factors after completely load shed of the critical bus.

Eigenvalue (Mvar / %)	Bus	Factor (-)
2.21654	26	0.167936
	27	0.16701
	12	0.154329
	28	0.14362
	13	0.13481
	24	0.066703
	11	0.052287
	10	0.035756
	9	0.032
	30	0.025058
	14	0.015675
	8	0.004652
4.465852	9	0.541871
	10	0.189382
	8	0.107956
	11	0.058957
	28	0.026436
	26	0.018823
	27	0.016214

	24	0.013057
	12	0.010143
	13	0.009922
	30	0.006235
10.206046	28	0.379561
	24	0.279088
	30	0.254547
	11	0.017955
	26	0.015515
	13	0.012181
	9	0.010358
	12	0.00765
	8	0.006431
	10	0.006318
	20	0.003871
11.039589	20	0.654281
	24	0.114903
	11	0.078371
	10	0.038417
	9	0.036138
	8	0.028376
	13	0.013025
	28	0.011786
	30	0.010754
	12	0.007439

Table A.4.11   The 23-bus system's branch participation factors after completely load shed of the critical bus.

Eigenvalue (Mvar / %)	Bus	Factor (-)
2.21654	54	1
	16	0.911049
	17	0.911049
	56	0.734862
	57	0.734862
	8	0.606019
	13	0.602183
	4	0.54574
	20	0.533142
	21	0.533142
	7	0.462748
	11	0.429363
	3	0.367655
	2	0.367655
	5	0.351716
	55	0.297198
	25	0.184387
	26	0.184387
	14	0.171189
	15	0.171189
	9	0.12838
	10	0.12838

	12	0.11907
	6	0.10843
	22	0.039709
	53	0.013663
	15	0.011572
	52	0.010617
4.465852	5	1
	4	0.980608
	2	0.759272
	3	0.759272
	8	0.58796
	6	0.414229
	7	0.20966
	54	0.182605
	16	0.147848
	17	0.147848
	56	0.13956
	57	0.13956
	20	0.10849
	21	0.10849
	13	0.094833
	55	0.080324
	11	0.036247
	25	0.026098
	26	0.026098

	14	0.023902
	15	0.023902
	12	0.011952
10.206046	57	1
	56	1
	54	0.996197
	20	0.927301
	21	0.927301
	55	0.835677
	16	0.636245
	17	0.636245
	5	0.418804
	3	0.269841
	2	0.269841
	4	0.220598
	7	0.158225
	8	0.132693
	13	0.130818
	11	0.082019
	22	0.052884
	25	0.04054
	26	0.04054
	6	0.03571
	53	0.035558
	9	0.035255

	10	0.035255
	52	0.032083
	12	0.020783
11.039589	5	1
	54	0.835584
	53	0.76071
	55	0.732558
	2	0.662139
	3	0.662139
	52	0.56552
	4	0.495986
	7	0.384729
	8	0.381673
	56	0.284149
	57	0.284149
	20	0.279511
	21	0.279511
	13	0.260634
	16	0.143676
	17	0.143676
	22	0.101496
	11	0.099936
	9	0.047463
	10	0.047463
	12	0.023978

	25	0.013527
	26	0.013527
	6	0.012251

Table A.4.12 The 23-bus system's right eigenvectors after completely load shed of the critical bus.

Eigenvalue (Mvar / %)	Bus	Right Eigenvector (-)
2.21654	8	0.070781
	7	0.000019
	9	0.186474
	10	0.19107
	11	0.227594
	16	0.001174
	15	0.009551
	14	0.124604
	13	0.365063
	12	0.390434
	27	0.406467
	28	0.379584
	30	0.16091
	18	0.000327
	26	0.408093
	20	0.004374
	22	0.007237
	24	0.262258
4.465852	8	0.331758
	7	0.000089
	9	0.74571
	10	0.427132

	11	0.234786
	16	0.001214
	15	0.004552
	14	-0.029946
	13	-0.096189
	12	-0.097231
	27	-0.123044
	28	-0.158282
	30	-0.078089
	18	-0.000159
	26	-0.132746
	20	-0.002604
	22	-0.003155
	24	-0.112838
10.206046	8	0.082651
	7	0.000022
	9	0.104995
	10	-0.079566
	11	-0.132029
	16	-0.000687
	15	-0.004475
	14	-0.059054
	13	-0.108683
	12	-0.086089
	27	-0.050742

	28	0.611042
	30	0.508154
	18	0.000955
	26	-0.122831
	20	-0.06181
	22	-0.016071
	24	-0.531334
11.039589	8	0.174068
	7	0.000047
	9	0.196576
	10	-0.196634
	11	-0.276502
	16	-0.001441
	15	-0.006703
	14	-0.067688
	13	-0.112798
	12	-0.085203
	27	-0.030953
	28	0.108052
	30	0.104768
	18	0.000998
	26	0.009447
	20	0.806219
	22	0.026545
	24	0.341956

Annex 4.4: The IEEE 14-bus system simulation results with considering various involved parameters in voltage stability.

Table A.4.13   The 14-bus system's V-Q self-sensitivities.

Bus	V-Q Sensitivity (% / Mvar)
14	0.223312
10	0.162144
7	0.141678
9	0.137696
12	0.137652
11	0.135261
13	0.08723
4	0.043989
5	0.042725

Table A.4.14   The 14-bus system's eigenvalues [Mvar / %]

2.079206
5.387656
7.598687
9.494185
16.098493

Table A.4.15   The 14-bus system's bus participation factors at different eigenvalues.

Eigenvalue (Mvar / %)	Bus	Factor (-)
2.079206	14	0.23739
	10	0.233253
	9	0.225559
	7	0.161638
	11	0.092557
	13	0.019842
	4	0.013906
	12	0.009483
	5	0.006371
5.387656	12	0.323579
	14	0.285885
	13	0.172902
	11	0.079953
	10	0.076486
	7	0.033499
	9	0.022179
	4	0.003595
	5	0.001922
7.598687	12	0.468034
	14	0.355491
	11	0.107117
	10	0.034797

	13	0.034373
9.494185	11	0.418265
	7	0.345085
	4	0.062215
	14	0.058422
	5	0.041866
	9	0.037012
	10	0.020534
	12	0.016393
16.098493	11	0.227307
	10	0.227216
	5	0.221857
	4	0.200844
	9	0.070958
	7	0.035629
	13	0.006621
	14	0.006493
	12	0.003075

Table A.4.16 The 14-bus system's branch participation factors

Eigenvalue (Mvar / %)	Name	Factor (-)
2.079206	BRANCH-10	1
	BRANCH-11	0.82639
	BRANCH-8	0.786916
	BRANCH-20	0.581746
	BRANCH-13	0.468115
	BRANCH-18	0.372482
	BRANCH-6	0.31722
	BRANCH-2	0.275689
	BRANCH-5	0.275512
	BRANCH-7	0.272732
	BRANCH-15	0.260207
	BRANCH-1	0.225952
	BRANCH-3	0.219702
	BRANCH-4	0.128633
	BRANCH-12	0.118457
	BRANCH-9	0.098463
	BRANCH-14	0.097689
	BRANCH-17	0.066393
	BRANCH-19	0.019637
	BRANCH-16	0.017989
5.387656	BRANCH-13	1
	BRANCH-11	0.596877
	BRANCH-12	0.369794

	BRANCH-8	0.283252
	BRANCH-10	0.180856
	BRANCH-15	0.15988
	BRANCH-6	0.137281
	BRANCH-20	0.1259
	BRANCH-17	0.110207
	BRANCH-7	0.081659
	BRANCH-5	0.080215
	BRANCH-3	0.069843
	BRANCH-19	0.052411
	BRANCH-14	0.036014
	BRANCH-16	0.026523
	BRANCH-4	0.023845
	BRANCH-2	0.006759
	BRANCH-9	0.005383
	BRANCH-1	0.003947
	BRANCH-18	0.002489
7.598687	BRANCH-20	1
	BRANCH-11	0.710742
	BRANCH-13	0.5639
	BRANCH-12	0.371587
	BRANCH-18	0.23176
	BRANCH-19	0.178634
	BRANCH-10	0.047346
	BRANCH-16	0.036154

	BRANCH-17	0.035537
	BRANCH-9	0.014825
	BRANCH-15	0.012819
	BRANCH-3	0.010127
	BRANCH-4	0.009024
	BRANCH-6	0.008007
	BRANCH-7	0.007872
	BRANCH-1	0.006118
	BRANCH-8	0.00438
	BRANCH-2	0.003734
	BRANCH-5	0.00235
	BRANCH-14	0.002136
9.494185	BRANCH-15	1
	BRANCH-11	0.88319
	BRANCH-18	0.550644
	BRANCH-8	0.375345
	BRANCH-6	0.359844
	BRANCH-3	0.292462
	BRANCH-5	0.24641
	BRANCH-4	0.20044
	BRANCH-1	0.17841
	BRANCH-20	0.166674
	BRANCH-14	0.078057
	BRANCH-10	0.06525
	BRANCH-17	0.053744

	BRANCH-2	0.052744
	BRANCH-7	0.048817
	BRANCH-12	0.039307
	BRANCH-16	0.034963
	BRANCH-19	0.03268
	BRANCH-13	0.016924
	BRANCH-9	0.009949
16.098493	BRANCH-15	1
	BRANCH-18	0.916244
	BRANCH-10	0.674298
	BRANCH-3	0.562022
	BRANCH-11	0.559835
	BRANCH-6	0.547825
	BRANCH-5	0.487285
	BRANCH-1	0.46463
	BRANCH-4	0.452405
	BRANCH-8	0.396007
	BRANCH-7	0.17606
	BRANCH-9	0.14582
	BRANCH-2	0.121711
	BRANCH-13	0.076532
	BRANCH-17	0.038263
	BRANCH-19	0.02839
	BRANCH-16	0.024881
	BRANCH-14	0.021924

	BRANCH-20	0.019327
	BRANCH-12	0.005569

Table A.4.17 The 14-bus system's generator participation factors.

Eigenvalue (Mvar / %)	Bus	Name	Factor (-)
2.079206	6	GN6 13.8	1
	2	GN2 69.0	0.339793
	3	GN3 69.0	0.203609
	1	GN1 69.0	0.102139
5.387656	6	GN6 13.8	1
	2	GN2 69.0	0.151969
	3	GN3 69.0	0.089806
	1	GN1 69.0	0.051791
7.598687	6	GN6 13.8	1
	2	GN2 69.0	0.006906
	3	GN3 69.0	0.004002
	1	GN1 69.0	0.002204
9.494185	2	GN2 69.0	1
	6	GN6 13.8	0.781699
	3	GN3 69.0	0.555697
	1	GN1 69.0	0.346538
16.098493	2	GN2 69.0	1
	6	GN6 13.8	0.925544
	3	GN3 69.0	0.491113
	1	GN1 69.0	0.387797

Table A.4.18   The 14-bus system's right eigenvectors

Eigenvalue (Mvar / %)	Bus	Right Eigenvector (-)
2.079206	4	0.119854
	5	0.080149
	12	0.096489
	13	0.138994
	7	0.401572
	9	0.476716
	10	0.48392
	14	0.48619
	11	0.302497
5.387656	4	-0.062035
	5	-0.045303
	12	0.56576
	13	0.414648
	7	-0.18428
	9	-0.150253
	10	-0.278529
	14	0.536672
	11	-0.282759
7.598687	4	0.004242
	5	0.003287
	12	0.681081
	13	0.185223
	7	0.011204

	9	0.006267
	10	0.188071
	14	-0.599086
	11	0.327647
9.494185	4	-0.254511
	5	-0.209542
	12	-0.126515
	13	0.014617
	7	-0.586949
	9	-0.192998
	10	0.143348
	14	0.241798
	11	0.64368
16.098493	4	0.451865
	5	0.478962
	12	-0.05637
	13	0.083066
	7	0.186375
	9	-0.265208
	10	-0.473655
	14	0.078179
	11	0.469899

CHAPTER 6

HIGHLIGHTS OF THE CHAPTERS

The lessons learned from Section 1.3 (Global Blackout Incidences) implies the tendency of increasing need for voltage stability analysis and mitigation of the system operation overall mismatches; because of a combination of events and system conditions, additional reactive power demand may lead to voltage collapse, causing a major blackout of a part or all of the system. It is extremely important to ensure the stability of the system in such situations. Voltage stability depends on how variations in reactive, as well as active power in the load area, affect the voltages at the load buses.

The voltage stability phenomenon customarily classified in steady state, quasi-steady state, dynamic, semi-dynamic, and transient analyzes. Whereas, from voltage instability occurrence time is classified in short-term, classical, and long-term voltage stability; and from types of disturbances aspect is considered into two categories small disturbance and large disturbances voltage stability, at literature. Just rely on either P-V or V-Q analysis for the purpose of the stability assessment is not enough to judge a power system condition and assesses the proximity to voltage collapse point. At the same time, coordination of both P-V and V-Q including P-Q analysis can be count a milestone of the results to be investigated. Consequently, Chapter 2 concludes that the indices can expose effective information related to voltage stability at any operating point, irrespective of the previous numerical limitation like Jacobian matrix's inverse unobtainability.

The result indicates that the improved indicator with considers the right eigenvector at the minimum eigenvalue can be counted an appropriate tool for weak bus identification and system voltage stability evaluation. In addition, the comparative analysis shows the consistency of the improved index with other indices, comparatively with less risk of uncertainty. Nevertheless, somewhat the proposed index is analogous to the $Q - V$ approach (sensitivity). Meanwhile, it has merit than the previous proposed indices at the same framework. At last, the collaboration of some aspects of voltage stability in this study can give a quick overview on the voltage stability assessment based on modal analysis.

Altogether, from Chapter 3, can be noted that the obtained information make evident the consistency of the studied indices with distinguish of the inconsistent indices in the way of the voltage stability analysis. The results are found that all indices in a category are coherent to theirs typical theoretical bases, and pursue the same performance. Accordingly, the results imply that the accordant voltage stability indices can be applied alternatively, except those are mentioned. The range of stability for most of the indices are between 0 and 1.0. Someway, it indicates that the indices discernment characteristics of performances are in accord. The study can either reveal that the indices ranking with respect to the worst node or area identification does not imply that an index drawback or bad ranking in compliance with other indices, it is useless. Whereas, each index is functional for a specific application. The simulation results negate the application of line indices in order to identify the critical bus in a power system. As well as, employing of the Jacobian matrix-based indices for online application due to theirs non-linearity properties at the collapse point and high computation time and space requirement. The voltage stability indices comparative analysis and the manifest study, can pave the ground for well application of the indices in various application of power system such as optimal placement of distributed generation, reactive power dispatch, and power management. Subsequently, the classification section is addressed that it is arduous to ascribe the voltage stability indices to static or dynamic classes.

The application of the results of the aforesaid chapters is discussed in Chapter 4 and 5 as following. The main scheme of this chapter was presented and illustrated. The shunt capacitors optimum placement is cheap from the economic point of view, and it is easy to use anywhere from technical point view. It can be implemented as a quick method to the assessment of weak bus and its feasibility to compensate and prevent voltage to going the unstable. The overall behavior of the test system is considered acceptable despite variation difference in sensitivity. The proposed method can be used, in addition of avoid voltage variation, in weak bus it might be used for extending the loadability limit for future demand.

It is found that the reactive power balance has been an important factor for the voltage stability phenomenon. The reactive power, customary is controlled by using shunt capacitors because from economic point view it is cost-effective and from technical point view it is easy to install and operate. This study is easily applicable for overall optimization of the stressed power system to optimize the weak bus which is experienced lack of reactive power in the system via quick and straightforward power flow simulation with various values for reactive power injection at different iterations.

Voltage stability problem has been a widespread concern in power systems as a result of heavier loadings. One procedure of avoiding voltage instability is to shed load of the critical buses/areas in the system. This chapter is conducted on a separate aspect of voltage stability; proposed assessment index load shedding via weak bus identification. The effectiveness of the proposed assessment index is tested on two relative test systems in normal and critical operations. The Jacobian matrix is simplified through singular value decomposition and Pseudo- inverse (linear algebra). The result indicates that the load shedding for the instability mitigation can be used as a useful tool. While, always rely on the load shedding may cause a bad situation for the utility company economic and reputation. Therefore, load shedding as a preventive tool for server instability condition to prevent voltage collapse is proposed.

REFERENCES

[1]   Liscouski, B., and W. Elliot. "Final report on the august 14, 2003 blackout in the united states and Canada: Causes and recommendations." A report to US Department of Energy 40, no. 4 (2004).

[2]   Doig Cardet, Christine Elizabeth. "Analysis on Voltage Stability Indices."

[3]   Steinmetz, Charles P. "Power control and stability of electric generating stations." American Institute of Electrical Engineers, Transactions of the 39, no. 2 (1920): 1215-1287.

[4]   Kundur, Prabha, John Paserba, Venkat Ajjarapu, Göran Andersson, Anjan Bose, Claudio Canizares, Nikos Hatziargyriou et al. "Definition and classification of power system stability IEEE/CIGRE joint task force on stability terms and definitions." Power Systems, IEEE Transactions on 19, no. 3 (2004): 1387-1401.

[5]   Kundur, Prabha. Power system stability and control. Edited by Neal J. Balu, and Mark G. Lauby. Vol. 7. New York: McGraw-hill, 1994.

[6]   Nakawiro, Worawat, and István Erlich. "Online voltage stability monitoring using artificial neural network." In Electric Utility Deregulation and Restructuring and Power Technologies, 2008. DRPT 2008. Third International Conference on, pp. 941-947. IEEE, 2008.

[7]   Power system security: web-based design tools. http://seniord.ece.iastate.edu/may0323/voltage_stab_def.html (accessed April, 2014).

[8]   Kearsley, Roger. "Restoration in Sweden and experience gained from the blackout of 1983." Power Systems, IEEE Transactions on 2, no. 2 (1987): 422-428.

[9]   Ohno, T., and S. Imai. "The 1987 tokyo blackout." In Power Systems

Conference and Exposition, 2006. PSCE'06. 2006 IEEE PES, pp. 314-318. IEEE, 2006.

[10]   Kurita, Atsushi, and T. Sakurai. "The power system failure on July 23, 1987 in Tokyo." In Decision and Control, 1988. Proceedings of the 27th IEEE Conference on, pp. 2093-2097. IEEE, 1988.

[11]   Machowski, Jan, Janusz Bialek, and Jim Bumby. Power system dynamics: stability and control. John Wiley & Sons, 2011.

[12]   Chebbo, A. M., M. R. Irving, and M. J. H. Sterling. "Voltage collapse proximity indicator: behaviour and implications." In IEE Proceedings C (Generation, Transmission and Distribution), vol. 139, no. 3, pp. 241-252. IEE, 1992.

[13]   Kessel, P., and H. Glavitsch. "Estimating the voltage stability of a power system." Power Delivery, IEEE Transactions on 1, no. 3 (1986): 346-354.

[14]   Tamura, Y., H. Mori, and S. Iwamoto. "Relationship between voltage instability and multiple load flow solutions in electric power systems." Power Apparatus and Systems, IEEE Transactions on 5 (1983): 1115-1125.

[15]   Kwatny, Harry G., A. Pasrija, and L. Bahar. "Static bifurcations in electric power networks: loss of steady-state stability and voltage collapse." Circuits and Systems, IEEE Transactions on 33, no. 10 (1986): 981-991.

[16]   Gao, Baofu, G. K. Morison, and Prabhashankar Kundur. "Voltage stability evaluation using modal analysis." Power Systems, IEEE Transactions on 7, no. 4 (1992): 1529-1542.

[17]   Electric Power Research Institute Technical Report, EPRI Power Systems Dynamics Tutorial. EPRI, Palo Alto, CA: 2009. 1016042.

[18]   Chakrabarti, Abhijit, et al. An introduction to reactive power control and voltage stability in power transmission systems. PHI Learning Pvt. Ltd., 2010.

[19]   FGL's 3 bus Test System. http://fglongatt.org/OLD/Test_Case_FGL's%203%20bus.html (accessed February, 2014).

[20]   Mansour, Yakout, Wilsun Xu, Fernando Alvarado, and Chhewang Rinzin. "SVC placement using critical modes of voltage instability." Power Systems, IEEE Transactions on 9, no. 2 (1994): 757-763.

[21]   Taylor, Carson W. Power system voltage stability. McGraw-Hill, 1994.

[22]   Momoh, James A., and Mohamed E. El-Hawary. Electric systems, dynamics, and stability with artificial intelligence applications. CRC Press, 1999.

[23]   Hill, David J., MK PAL, XU WILSUN, YAKOUT MANSOUR, CO NWANKPA, L. Xu, and R. Fischl. "Nonlinear dynamic load models with recovery for voltage stability studies. Discussion. Authors' response." IEEE Transactions on Power Systems 8, no. 1 (1993): 166-176.

[24]   Ajjarapu, Venkataramana. Computational techniques for voltage stability assessment and control. Springer, 2007.

[25]   Li, Wenyuan. Probabilistic transmission system planning. Vol. 65. John Wiley & Sons, 2011.

[26]   Danish, Mir Sayed Shah, Atsushi Yona, and Tomonobu Senjyu. "Voltage stability assessment index for recognition of proper bus for load shedding. "Information Science, Electronics and Electrical Engineering (ISEEE), 2014 International Conference on, vol. 1, pp. 636-639. IEEE, 2014.

[27]   Miljkovic, Sladjana, Marko Miladinovic, Predrag Stanimirovic, and Igor Stojanovic. "Application of the pseudoinverse computation in reconstruction of blurred images." Filomat 26, no. 3 (2012): 453-465.

[28]   Calderon-Guizar, J. G., and E. F. Noriega. "Speeding up the computations of the minimum singular value of the load flow Jacobian matrix." Power Engineering Review, IEEE 19, no. 9 (1999): 55-56.

[29]   Chang, Po Rong, and C. S. G. Lee. "Residue arithmetic VLSI array architecture for manipulator pseudo-inverse Jacobian computation." Robotics and Automation, IEEE Transactions on 5, no. 5 (1989): 569-582.

[30]   Prada, R. B., and J. O. R. Dos Santos. "Fast nodal assessment of static voltage stability including contingency analysis." Electric power

systems research 51, no. 1 (1999): 55-59.

[31] Su, Yongchun, Shijie Cheng, Jinyu Wen, and Yonggao Zhang. "Reactive power generation management for the improvement of power system voltage stability margin." In Intelligent Control and Automation, 2006. WCICA 2006. The Sixth World Congress on, vol. 2, pp. 7466-7469. IEEE, 2006.

[32] Berizzi, A., P. Bresesti, P. Marannino, G. Granelli, and M. Montagna. "System-area operating margin assessment and security enhancement against voltage collapse." Power Systems, IEEE Transactions on 11, no. 3 (1996): 1451-1462.

[33] Barquin, J. "On voltage stability regions and voltage secondary control." In Electric Power Engineering, 1999. PowerTech Budapest 99. International Conference on, p. 246. IEEE, 1999.

[34] Zambroni de Souza, A. C. "Discussion on some voltage collapse indices. "Electric Power Systems Research 53, no. 1 (2000): 53-58.

[35] Sauer, P. W., and M. A. Pai. "Power system steady-state stability and the load-flow Jacobian." Power Systems, IEEE Transactions on 5, no. 4 (1990): 1374-1383.

[36] Lof, P-A., G. Andersson, and D. J. Hill. "Voltage stability indices for stressed power systems." Power Systems, IEEE Transactions on 8, no. 1 (1993): 326-335.

[37] Tiranuchit, A., and R. J. Thomas. "A posturing strategy against voltage instabilities in electric power systems." Power Systems, IEEE Transactions on3, no. 1 (1988): 87-93.

[38] Hong, Young-Huei, Ching-Tsai Pan, and Wen-Wei Lin. "Fast calculation of a voltage stability index of power systems." Power Systems, IEEE Transactions on 12, no. 4 (1997): 1555-1560.

[39] Amer, AL-Hinai. "Voltage Collapse Prediction for Interconnected Power Systems." PhD diss., West Virginia University, 2000.

[40] Tiranuchit, A., L. M. Ewerbring, R. A. Duryea, R. J. Thomas, and F. T. Luk. "Towards a computationally feasible on-line voltage instability index." Power Systems, IEEE Transactions on 3, no. 2 (1988): 669-675.

[41] Klema, Virginia, and Alan J. Laub. "The singular value decomposition:

Its computation and some applications." Automatic Control, IEEE Transactions on25, no. 2 (1980): 164-176.

[42]   Padiyar, K.R. Power System Dynamics: Stability and Control. BS Publications., 2008.

[43]   Sinha, A. K., and D. Hazarika. "A comparative study of voltage stability indices in a power system." International journal of electrical power & energy systems22, no. 8 (2000): 589-596.

[44]   Sun, Huadong, Xiaoxin Zhou, and Ruomei Li. "Accuracy analysis of static voltage stability indices based on power flow model." In Transmission and Distribution Conference and Exhibition: Asia and Pacific, 2005 IEEE/PES, pp. 1-7. IEEE, 2005.

[45]   Karbalaei, F., H. Soleymani, and S. Afsharnia. "A comparison of voltage collapse proximity indicators." In IPEC, 2010 Conference Proceedings, pp. 429-432. IEEE, 2010.

[46]   Esmaili, Masoud, Esmail Chaktan Firozjaee, and Heidar Ali Shayanfar. "Optimal placement of distributed generations considering voltage stability and power losses with observing voltage-related constraints." Applied Energy 113 (2014): 1252-1260.

[47]   Danish, Mir Sayed Shah, A. Yona, and T. Senjyu. "Optimum loadability improvement of weak buses using shunt capacitors to enhance voltage stability margin." International Conference on Engineering and Applied Science: ICEAS, pp. 1063-1069. 2013.

[48]   Yao, Zhang, and Song Wennan. "Power system static voltage stability limit and the identification of weak bus." In TENCON'93. Proceedings. Computer, Communication, Control and Power Engineering. 1993 IEEE Region 10 Conference on, pp. 157-160. IEEE, 1993.

[49]   Venikov, V. A., V. A. Stroev, V. I. Idelchick, and V. I. Tarasov. "Estimation of electrical power system steady-state stability in load flow calculations." Power Apparatus and Systems, IEEE Transactions on 94, no. 3 (1975): 1034-1041.

[50]   Eremia, Mircea, and Mohammad Shahidehpour, eds. Handbook of Electrical Power System Dynamics: Modeling, Stability, and Control. Vol. 92. John Wiley & Sons, 2013.

[51] Grafarend, Erik W., Béla Paláncz, and Piroska Zaletnyik. Algebraic geodesy and geoinformatics. Springer Berlin Heidelberg, 2010.

[52] Stewart, Gilbert W. Matrix Algorithms Volume 1: Eigensystems. Vol. 1. Siam, 1998.

[53] Levine, William S., ed. Control System Applications. CRC press, 2011.

[54] Golub, Gene, and William Kahan. "Calculating the singular values and pseudo-inverse of a matrix." Journal of the Society for Industrial & Applied Mathematics, Series B: Numerical Analysis 2, no. 2 (1965): 205-224.

[55] NEPLAN® is a high-end power system analysis tool for applications in transmission, distribution, generation, industrial, renewable energy systems, and Smart Grid application.http://www.neplan.ch. (Accessed, December, 2013).

[56] Reis, Claudia, and FP Maciel Barbosa. "A comparison of voltage stability indices." In Electrotechnical Conference, 2006. MELECON 2006. IEEE Mediterranean, pp. 1007-1010. IEEE, 2006.

[57] Gao, Peng, Libao Shi, Liangzhong Yao, Yixin Ni, and Masoud Bazargan. "Multi-criteria integrated voltage stability index for weak buses identification." In Transmission & Distribution Conference & Exposition: Asia and Pacific, 2009, pp. 1-5. IEEE, 2009.

[58] Van Cutsem, Thierry. "A method to compute reactive power margins with respect to voltage collapse." Power Systems, IEEE Transactions on 6, no. 1 (1991): 145-156.

[59] De, Mala, and Swapan K. Goswami. "Optimal Reactive Power Procurement With Voltage Stability Consideration in Deregulated Power System." Power Systems, IEEE Transactions on 29, no. 5 (2014): 1-9.

[60] Zhang, Zhihua. "Pseudo-inverse multivariate/matrix-variate distributions." Journal of Multivariate Analysis 98, no. 8 (2007): 1684-1692.

[61] Balamourougan, V., T. S. Sidhu, and M. S. Sachdev. "Technique for online prediction of voltage collapse." In Generation, Transmission and Distribution, IEE Proceedings-, vol. 151, no. 4, pp. 453-460. IET, 2004.

[62] Power System Test Case Archive – UWEE. http://www.ee.washington.edu/research/pstca (accessed February 12, 2014).

[63] Chiang, Hsiao-Dong. Direct methods for stability analysis of electric power systems: theoretical foundation, BCU methodologies, and applications. John Wiley & Sons, 2011.

[64] Massucco, S., S. Grillo, A. Pitto, and F. Silvestro. "Evaluation of some indices for voltage stability assessment." In PowerTech, 2009 IEEE Bucharest, pp. 1-8. IEEE, 2009.

[65] Suganyadevia, M. V., and C. K. Babulal. "Estimating of loadability margin of a power system by comparing Voltage Stability Indices." In Control, Automation, Communication and Energy Conservation, 2009. INCACEC 2009. 2009 International Conference on, pp. 1-4. IEEE, 2009.

[66] Cupelli, Marco, Christine Doig Cardet, and Antonello Monti. "Comparison of line voltage stability indices using dynamic real time simulation." In Innovative Smart Grid Technologies (ISGT Europe), 2012 3rd IEEE PES International Conference and Exhibition on, pp. 1-8. IEEE, 2012.

[67] Cupelli, M., C. Doig Cardet, and A. Monti. "Voltage stability indices comparison on the IEEE-39 bus system using RTDS." In Power System Technology (POWERCON), 2012 IEEE International Conference on, pp. 1-6. IEEE, 2012.

[68] Power System Dynamic Performance Committee: Power System Stab Subcommittee. Voltage stability assessment: concepts, practices and tools. In: Claudio Cafiizares, editor. Voltage stability indices, IEEE Power & Energy Soc; 2013, p. 172–241.

[69] Nizam, Muhammad, Azah Mohamed, and Aini Hussain. "Performance evaluation of voltage stability indices for dynamic voltage collapse prediction. "Journal of Applied Sciences 6, no. 5 (2006): 1104-1113.

[70] Acharya, N. Vishwas, and P. S. Rao. "A new voltage stability index based on the tangent vector of the power flow jacobian." In Innovative Smart Grid Technologies-Asia (ISGT Asia), 2013 IEEE, pp. 1-6. IEEE,

2013.

[71]   Wang, Yang, Caisheng Wang, Feng Lin, Wenyuan Li, Le Yi Wang, and Junhui Zhao. "Incorporating Generator Equivalent Model Into Voltage Stability Analysis." (2013): 1-10.

[72]   Pérez-Londoño, S., L. F. Rodríguez, and G. Olivar. "A Simplified Voltage Stability Index (SVSI)." International Journal of Electrical Power & Energy Systems 63 (2014): 806-813.

[73]   Kayal, Partha, and C. K. Chanda. "Placement of wind and solar based DGs in distribution system for power loss minimization and voltage stability improvement." International Journal of Electrical Power & Energy Systems 53 (2013): 795-809.

[74]   Zabaiou, Tarik, Louis-A. Dessaint, and Innocent Kamwa. "Preventive control approach for voltage stability improvement using voltage stability constrained optimal power flow based on static line voltage stability indices." IET Generation, Transmission & Distribution 8, no. 5 (2014): 924-934.

[75]   Moghavvemi, M., and O. Faruque. "Real-time contingency evaluation and ranking technique." In Generation, Transmission and Distribution, IEE Proceedings-, vol. 145, no. 5, pp. 517-524. IET, 1998.

[76]   Musirin, Ismail, and Titik Khawa Abdul Rahman. "On-line voltage stability based contingency ranking using fast voltage stability index (FVSI)." InTransmission and Distribution Conference and Exhibition 2002: Asia Pacific. IEEE/PES, vol. 2, pp. 1118-1123. IEEE, 2002.

[77]   Moghavvemi, Mahmoud, and G. B. Jasmon. "New method for indicating voltage stability condition in power system." In IEE Int. Conf. an Power Engineering. 1997.

[78]   Musirin, Ismail, and TK Abdul Rahman. "Estimating maximum loadability for weak bus identification using FVSI." IEEE power engineering review 22, no. 11 (2002): 50-52.

[79]   Mohamed, A., G. B. Jasmon, and S. Yusoff. "A static voltage collapse indicator using line stability factors." Journal of industrial technology 7, no. 1 (1989): 73-85.

[80]   Jasmon, G. B., and L. H. C. C. Lee. "New contingency ranking

technique incorporating a voltage stability criterion." In Generation, Transmission and Distribution, IEE Proceedings C, vol. 140, no. 2, pp. 87-90. IET, 1993.

[81]   Naishan, Hang, Xu Tao, and Liao Qinghua. "The analysis of abundance index of voltage stability based circuit theory." Guangxi Electric Power 2 (2006): 12-14.

[82]   Aman, M. M., G. B. Jasmon, H. Mokhlis, and A. H. A. Bakar. "Optimal placement and sizing of a DG based on a new power stability index and line losses." International Journal of Electrical Power & Energy Systems 43, no. 1 (2012): 1296-1304.

[83]   Indices predicting voltage collapse including dynamic phenomena. CIGRE, 1994.

[84]   Overbye, Thomas J., and Christopha L. DeMarco. "Improved techniques for power system voltage stability assessment using energy methods." Power Systems, IEEE Transactions on 6, no. 4 (1991): 1446-1452.

[85]   Ajjarapu, Venkataramana, and Colin Christy. "The continuation power flow: a tool for steady state voltage stability analysis." Power Systems, IEEE Transactions on 7, no. 1 (1992): 416-423.

[86]   Glavic, Mevludin, and Thierry Van Cutsem. "Wide-area detection of voltage instability from synchronized phasor measurements. Part I: Principle." Power Systems, IEEE Transactions on 24, no. 3 (2009): 1408-1416.

[87]   Yang, Chien-Feng, Gordon G. Lai, Chia-Hau Lee, Ching-Tzong Su, and Gary W. Chang. "Optimal setting of reactive compensation devices with an improved voltage stability index for voltage stability enhancement." International Journal of Electrical Power & Energy Systems 37, no. 1 (2012): 50-57.

[88]   Mohamed, A., and G. B. Jasmon. "Voltage contingency selection technique for security assessment." In IEE Proceedings C (Generation, Transmission and Distribution), vol. 136, no. 1, pp. 24-28. IEE, 1989.

[89]   Chakravorty, M., and D. Das. "Voltage stability analysis of radial distribution networks." International Journal of Electrical Power &

Energy Systems 23, no. 2 (2001): 129-135.

[90]   Haque, M. H. "Use of local information to determine the distance to voltage collapse." In Power Engineering Conference, 2007. IPEC 2007. International, pp. 407-412. IEEE, 2007.

[91]   Wang, Yang, Wenyuan Li, and Jiping Lu. "A new node voltage stability index based on local voltage phasors." Electric Power Systems Research 79, no. 1 (2009): 265-271.

[92]   De Souza, AC Zambroni, Claudio A. Canizares, and Victor H. Quintana. "New techniques to speed up voltage collapse computations using tangent vectors. "Power Systems, IEEE Transactions on 12, no. 3 (1997): 1380-1387.

[93]   Chiang, Hsiao-Dong, and Rene Jean-Jumeau. "Toward a practical performance index for predicting voltage collapse in electric power systems." Power Systems, IEEE Transactions on 10, no. 2 (1995): 584-592.

[94]   Berizzi, A., P. Finazzi, D. Dosi, P. Marannino, and S. Corsi. "First and second order methods for voltage collapse assessment and security enhancement. "Power Systems, IEEE Transactions on 13, no. 2 (1998): 543-551.

[95]   Li, W., Y. Wang, and T. Chen. "Investigation on the Thevenin equivalent parameters for online estimation of maximum power transfer limits." IET generation, transmission & distribution 4, no. 10 (2010): 1180-1187.

[96]   Julian, D. E., R. P. Schulz, K. T. Vu, W. H. Quaintance, N. B. Bhatt, and D. Novosel. "Quantifying proximity to voltage collapse using the voltage instability predictor (VIP)." In Power Engineering Society Summer Meeting, 2000. IEEE, vol. 2, pp. 931-936. IEEE, 2000.

[97]   Vu, Khoi, Miroslav M. Begovic, Damir Novosel, and Murari Mohan Saha. "Use of local measurements to estimate voltage-stability margin." Power Systems, IEEE Transactions on 14, no. 3 (1999): 1029-1035.

[98]   Milosevic, Borka, and Miroslav Begovic. "Voltage-stability protection and control using a wide-area network of phasor measurements." Power Systems, IEEE Transactions on 18, no. 1 (2003): 121-127.

[99]   Gong, Yanfeng, Noel Schulz, and Armando Guzman. "Synchrophasor-based real-time voltage stability index." In Power Systems Conference and Exposition, 2006. PSCE'06. 2006 IEEE PES, pp. 1029-1036. IEEE, 2006.

[100]   IEEE Power Engineering Society. Conference papers. USA: IEEE, University of Michigan; 1995.

[101]   PowerWorld Corporation. http://www.powerworld.com (accessed March 1, 2010).

[102]   A MATPOWER Power System Simulation Package. http://www.pserc.cornell.edu/matpower (accessed August 12, 2014).

[103]   Das, D., D. P. Kothari, and A. Kalam. "Simple and efficient method for load flow solution of radial distribution networks." International Journal of Electrical Power & Energy Systems 17, no. 5 (1995): 335-346.

[104]   Moghavvemi, M., and F. M. Omar. "Technique for contingency monitoring and voltage collapse prediction." IEE Proceedings-Generation, Transmission and Distribution 145, no. 6 (1998): 634-640.

[105]   Zambroni de Souza, A. C. "Tangent vector applied to voltage collapse and loss sensitivity studies." Electric power systems research 47, no. 1 (1998): 65-70.

[106]   Dobson, Ian, and Liming Lu. "Voltage collapse precipitated by the immediate change in stability when generator reactive power limits are encountered. "Circuits and Systems I: Fundamental Theory and Applications, IEEE Transactions on 39, no. 9 (1992): 762-766.

[107]   Omidi, H., B. Mozafari, A. Parastar, and M. A. Khaburi. "Voltage stability margin improvement using shunt capacitors and active and reactive power management." In Electrical Power & Energy Conference (EPEC), 2009 IEEE, pp. 1-5. IEEE, 2009.

[108]   Kaur, Damanjeet, and Jaydev Sharma. "Multiperiod shunt capacitor allocation in radial distribution systems." International Journal of Electrical Power & Energy Systems 52 (2013): 247-253.

[109]   Prada, R. B., E. G. C. Palomino, L. A. S. Pilotto, and A. Bianco. "Weakest bus, most loaded transmission path and critical branch

identification for voltage security reinforcement." Electric power systems research 73, no. 2 (2005): 217-226.

[110] Arunagiri, A., and B. Venkatesh. "Simulation of voltage stability and alleviation through knowledge based system." American Journal of Applied Sciences 1, no. 4 (2004): 354-357.

[111] Kumaraswamy, I., W. V. Jahnavi, T. Devaraju, and P. Ramesh. "An Optimal Power Flow (OPF) Method with Improved Voltage Stability Analysis." In Proceedings of the World Congress on Engineering, vol. 2, pp. 4-6. 2012.

[112] Van Cutsem, Thierry, and Costas Vournas. Voltage stability of electric power systems. Vol. 441. Springer, 1998.

[113] Chattopadhyay, D., and B. B. Chakrabarti. "Reactive power planning incorporating voltage stability." International journal of electrical power & energy systems 24, no. 3 (2002): 185-200.

[114] Alizadeh Mousavi, Omid, Mokhtar Bozorg, and Rachid Cherkaoui. "Preventive reactive power management for improving voltage stability margin." Electric Power Systems Research 96 (2013): 36-46.

[115] RAJ, PV, and M. Sudhakaran. "Optimum load shedding in power system strategies with voltage stability indicators." Engineering 2, no. 01 (2010): 12.

[116] Begovic, Miroslav M., and Arun G. Phadke. "Control of voltage stability using sensitivity analysis." Power Systems, IEEE Transactions on 7, no. 1 (1992): 114-123.

[117] Nagao, T., K. Tanaka, and K. Takenaka. "Development of static and simulation programs for voltage stability studies of bulk power system." Power Systems, IEEE Transactions on 12, no. 1 (1997): 273-281.

[118] Iba, Kenji, Hiroshi Suzuki, Masanao Egawa, and Tsutomu Watanabe. "Calculation of critical loading condition with nose curve using homotopy continuation method." Power Systems, IEEE Transactions on 6, no. 2 (1991): 584-593.

[119] Reis, Claudia, Antonio Andrade, and F. P. Maciel. "Line stability indices for voltage collapse prediction." In Power Engineering, Energy and Electrical Drives, 2009. POWERENG'09. International

Conference on, pp. 239-243. IEEE, 2009.

[120] Aly, Mohamed, and Mamdouh Abdel-Akher. "A continuation power-flow for distribution systems voltage stability analysis." In Kota Kinabalu. 2012 IEEE International Conference on Power and Energy (PECon), pp. 470-475. 2012.

INDEX

ABOUT THE AUTHOR

DANISH Mir Sayed Shah had received the BSc. in Electrical & Electronics Engineering from Kabul University in 2009; and expected to achieve the MSc. in Electrical Engineering from Ryukyus University on March 2015. He is a graduate student member of IEEJ, IEEE, IET and many others professional associations. As an active researcher, in addition of several academic publications, asked to be Editor for several international journals as well. He had worked in different positions such as electrical design engineer, urban electric power system planner, team leader, lecturer, educational field manager, project coordinator, and electrical engineer in biometric and datacenter technologies with different national and international organizations between 2004 and 2012. During this period, his contributions are constantly appreciated by awards. He is mostly interested in renewable energy, power system voltage stability, power system optimization, load shedding, power management, smart house and smart grid technologies, biometric intelligent system, and M2M technology.